"Technological evolution demands more than just adoption; it requires visionary leadership. *Working with Dinosaurs* empowers executives to become true catalysts for change within their organizations." —**BONITA STEWART**, Co-Founder and Managing Partner, BAG Ventures, Former Google VP

"If you still think tech is 'someone else's job,' read this before your competitors do. East's real-world war stories and simple frameworks make the future feel less like science fiction and more like your next agenda item." —**MARIO CIABARRA**, CEO & Founder, QuantumMetric

"Leading technological change at the executive level requires a unique blend of vision and pragmatism. *Working with Dinosaurs* nails it, offering practical insights that leaders can implement immediately to drive real results." —**BRIAN P. CHADWICK**, Esq. Chief Legal Officer at Devolver Digital, Inc.

"This book understands the unique pressures and complexities faced by C-suite leaders in driving technological change within established companies. It's a survival guide for the digital age, cleverly framed and powerfully insightful." —**ALEKSANDAR CABRILO**, Founder & CEO, HTEC Group

Working *With* Dinosaurs

HOW TO LEAD TECHNOLOGICAL
EVOLUTION FROM THE C-SUITE

Marcus East

WREN HOUSE
press

WORKING WITH DINOSAURS
How to Lead Technological Evolution from the C-Suite
First Edition

ISBN   978-1-967115-14-3   *Hardcover*
       978-1-967115-13-6   *Paperback*
       978-1-967115-12-9   *Ebook*
       978-1-967115-15-0   *Audiobook*

LCCN 2025914032

Contents

Introduction

You may be wondering what a book about IT departments and digital transformation has to do with dinosaurs.

Well, you already know what happened to the dinosaurs. They disappeared because they could not adapt to a rapidly changing environment.

If you are a company CEO or a tech leader, you can probably relate to that situation. Everything changes so rapidly in our digital world. Perhaps you, or your organization, have already been left behind. Perhaps your concerns are also personal: Perhaps you're worried that you (or your job position) are in danger of extinction too.

Let me explain why I have called this book *Working with Dinosaurs*.

We find ourselves working with technological dinosaurs every day: within our businesses, our corporations, our organizations.

We have to deal with accounting dinosaurs, financial dinosaurs, structural dinosaurs. Our computing systems are often outdated by the time they are installed, and our software too.

Most challenging are the human dinosaurs we deal with—especially those on our tech teams, but often in management as well. Infrastructure can be switched out; software can be rewritten. But humans tend to resist change and may take some effort to be convinced.

Whether we like it or not, our world is changing—and at an ever-accelerating rate. We had around a half century to transition from installing the first mainframe computers to beginning to base all our business operations on the internet. But it took us only a little over a decade to consider smartphones and ubiquitous connectivity both necessary and normal. Around the same time, the power of cloud computing was already starting to be harnessed by the corporate world.

Even amid all that change, though, we still felt grounded in our traditional roles as executives; our job descriptions were quite well defined by our titles. The CFO handled finance, the CMO looked after the marketing, the COO managed operations, and the CEO oversaw everything and planned for the future. The CIO (if there was one) was in charge of computing functions—overseeing both the hardware/infrastructure as well as programming and coding and taking on specific computing projects as requested by the business end of the organization.

These days, my friends, things are completely different! As the technological changes seem to appear weekly, or even daily, we find ourselves in a technological arms race, trying to process an apparently endless flow of press releases announcing new breakthroughs in AI or robotics or genetic manipulation.

This means that technology is now thoroughly integrated with every aspect of the business world: from finance to the processes on the manufacturing floor, to transport, to marketing and product development. There is no longer any place in management for anyone who is not conversant with modern technologies.

Thus, the "tech" role can no longer be siloed: Every executive and every board member must also be a "tech person."

Traditionally, the IT leader (the CIO, or in recent years more often the CTO or CDO) took care of the computing systems. They cared for the computers, developed software, and managed their team of technologists. They were the person who gave a "yea or nay" to projects proposed by management.

But with cloud computing, we no longer need those engineers down in the basement, plugging in cables and writing code. We can contract those services out. (If your company has not already harnessed the power of the cloud, then do read on!)

We still need our tech leaders, to be sure, but they no longer need to be squirrelled away, looking downward. They need to be looking upward and outward.

Of course, an IT leader still needs to be a technology special-ist! But they also need to be creative and a dreamer. They need to be capable of developing a vision of how new technologies can be used to innovate and drive change in an organization. They need to work closely with the business end of the organization in envisioning a path into the future. And they need to be a com-municator, able to inspire their team to come on board and buy in to this new vision. This is a much different role from the CIO of yore.

I came to think this way over the course of my own journey in the tech world, beginning in the mid-1990s. I first worked with IBM, and then took on various executive positions (CIO or CTO) with several different companies that were trying to keep pace with our changing technological world. My first experience leading a substantial digital transformation for an organization was from 2009 to 2011, when I was CTO and head of Future Media and Technology for the UK-based char-ity Comic Relief.

This was the first of many projects over the next decade and a half—working for companies including Apple, Marks and Spencer, National Geographic, Google, and T-Mobile—that showed me how much the IT leader's role has changed and must continue to evolve. Seeing this evolution play out on such a range of projects has utterly changed my thinking about what, actually, the IT leader of *today* needs to do: The position of the new IT leader is still about technology. But even more, now it is also about *leading*.

This shift has profound implications for your organization. First, it means that your IT personnel will take on different roles. By outsourcing fundamental computing operations to the cloud, your IT personnel can now participate in the vision for the future. They can focus on creating products and experiences that create direct value for your organization.

Second, those changed roles and responsibilities will mandate an entirely new way to structure your IT department. Most significantly, this means completely rethinking how your entire organization relates to and integrates with IT.

Ditching your dinosaurs means much more than switching out your old computers for new ones, and maybe upskilling some of your people. The powers of the cloud, and the rapidly growing capabilities of AI, require a complete re-envisioning of the structure of your business organization. This means changing roles and capabilities, and also changing relationships.

With this book, I aim to show you how to do that.

PART I

Adapt

ONE

Dinosaurs *in the* Driving Seat

W E TOUCHED ON THE PARALLELS BETWEEN today's business leaders and yesteryear's dinosaurs in the introduction, but to fully understand my choice of metaphor, it helps to understand the *full* story of the dinosaurs.

It's not an uplifting story at first glance; in fact, it ends with the dinosaurs disappearing—a final act if ever there was one. But let's look a little closer.

Yes, it's true that some dinosaurs did die out, their genetic lineages extinguished forever. But there's another side to this story: Other dinosaurs were agile and adaptable. They reacted rapidly to the abrupt and unexpected changes in their environment. Those dinosaurs managed to survive the Cretaceous meteor impact and flourish. In fact, those highly evolved and successful dinosaurs are still with us today.

FROM CRISIS, OPPORTUNITY

Long before you picked up this book, I'm sure you were already familiar with the meteor impact that initiated the great global extinction event at the end of the Cretaceous period.

This was only one of many mass extinction events that have occurred throughout geological time. Periods of rapid environmental change, due to meteor impacts or volcanic eruptions so massive that they affected our planet's atmosphere, have long led to hardship for some species, and even to their extinction.

For just as long, however, challenging and rapidly changing conditions have also done exactly the opposite. Rather than initiating death and destruction, they have instigated rapid evolution and innovation. Consider that previous mass extinction events, hundreds of millions of years before the Cretaceous meteor impact, were what set the scene for the emergence of the dinosaurs in the first place. The outcome of such events depends upon who is ready to adapt.

For example, some 185 million years *before* the Cretaceous meteor slammed into our planet came a series of immense volcanic eruptions in the region that is now Siberia. These eruptions elevated levels of CO_2, methane, and sulfur dioxide, leading to global warming. They also induced ocean acidification and reduced the amount of oxygen in the atmosphere to levels too low for most life to survive.

This was the largest global extinction event ever to have occurred on this planet, and it was aptly named the "Great Dying." By the end of the Great Dying, over 96 percent of marine species, over 70 percent of land animals, and most plants had gone extinct. For example, the changed environmental conditions of the Great Dying were hard on the synapsids, one line of primitive creatures that were ancestors of the mammals.

However, as always, from crisis comes opportunity.

The sauropsids, a line of lizard-like creatures, took advantage of the changed conditions. They quickly moved into the ecological niches vacated by the synapsids and flourished in their new and expanded environments. Some became herbivores, some became predators. Some developed crocodilian armor to protect themselves from predators, while others developed long necks for browsing shrubbery. Some learned to run on their hind legs and developed sharp teeth for catching prey, while others became small and fleet-footed so they could escape.

In other words, the sauropsids changed their behaviors and evolved to exploit different resources. These new, evolved sauropsids flourished, expanding in range and population, marking the beginning of the dinosaur age.

Within the short geological time of a few million years, the Permian Great Dying event was over, and we were well into the Triassic period. Life was fairly stable for millions of years, through the Jurassic and Cretaceous—until that most unforeseen of events: An asteroid ten kilometers wide smashed into the earth. It generated giant heat waves and fires, hurled tsunami waves over much of the American continents, and threw smoke and soot high into the atmosphere, plunging the planet into years of darkness.

This time, it was the synapsids' turn to shine. These mammal ancestors had spent millions of years evolving so they could avoid being eaten! Because of these adaptations, they were now able to exploit the niches vacated by the dinosaurs, exploding into a number of species and expanding over the planet. Some became smaller and lighter and more fleet-footed—like the mammals before them had done. Others evolved to develop fluffy quilled coverings to hold their heat in—what we now call feathers—and even learned to generate their own heat. Those agile and adaptive dinosaurs evolved into the birds—one of the most successful animal families, now with representatives in almost every ecological niche on our planet. A true story of a successful multinational! The ways in which some prehistoric animals successfully adapted and evolved can act as inspiration for the leaders of today.

MORE METEORS: MORE OPPORTUNITIES TO EVOLVE

I am telling this story of evolution and adaptation because this metaphor applies so well to the business world—the business climate, much like the physical climate, has drastically changed again and again over time.

When the world is stable and change is slow, business leaders can get away with doing what they have always done and get along just fine. That's how the dinosaurs managed to dominate our planet for 150 million years.

But when the world suddenly changes around us, it is those who are agile, able to respond and adapt, who will flourish. Those who cannot will be relegated to the background or die out altogether. (And beware, there are always lurkers in the background, just like those little mammals surviving in their burrows for millions of years, waiting to pounce and take over any new niche that opens up.)

To put it simply: Earth-shattering events that change the business climate require us to adapt and evolve, or we risk extinction.

Alternatively, we can adapt and thrive. Some businesses, organizations, and multinationals have weathered these historic storms successfully. Consider Fortnum & Mason, Lloyd's of London, The Royal Mint, National Geographic, Nakamura Shaji, and Nintendo. They have all been hit by numerous meteoric events, and they have evolved and thus survived—albeit in most cases (like the birds) not in the same form as the original.

One thing that won't happen, though—the meteors won't stop coming. They may even become more frequent. Witness the many meteors that have impacted the business world in the last half century, all related to evolution in technology. (Not that this is a recent phenomenon; I would argue that one of the first technological events to impact humankind occurred many centuries ago: the invention of writing. But let's bypass that for now, shall we?)

In the next several sections, we will look at recent meteor impacts, especially those related specifically to digital technology and to the appearance of IT departments as integral parts of business organizations.

The Mainframe Era

Remember mainframes? Maybe not—the first mainframe computer was built in 1943; they first became available to businesses in the late 1950s. Nonetheless, mainframe computing was the meteor impact that initiated the computer age, and along with it the formation of what we now call IT departments.

Back then, they were often referred to as the data processing departments—which is mainly what they did. Mainframes took over doing the heavy processing work: processing transactions for financial services organizations or doing stock updates for retailers and distributors. The IT team focused mainly on data processing. They ensured that their systems were available to the users upstairs through monochrome "dummy" terminals, which were no more than data entry screens for the centralized computers.

By the mid-1960s, minicomputers also appeared on the scene. They were much smaller than mainframes, taking up the space of a few cabinets rather than an entire room. However, they were powerful for their time, taking advantage of transistors and core memory technologies, and usually dedicated to specific tasks such as accounting or data processing.

Adoption of mainframe computers and then minicomputers by corporations represents the first major technological meteor to impact how business was conducted in our world. But before long, another meteor would appear, rocking this new techno-logical world.

The Personal Computer Revolution

The challenge to the era of mainframe and minicomputer computing came in the late 1970s, with the appearance of per-sonal computers, or PCs. Astonishingly, these tiny but pow-erful machines didn't require entire climate-regulated rooms. Suddenly, your computer could fit on your desk—and soon, into your briefcase or backpack.

A struggle ensued, with companies like Apple, Commodore, and Tandy juggling for position, trying different adaptations, and searching out new niches, behaving much like synapsids and sauropsids in the period right after the Great Dying.

Almost immediately, a new player emerged. IBM, long a dominating force in mainframe computing, launched their IBM Personal Computer in 1981. With this, personal computing systems gained traction in business, forever changing the nature of enterprise computing.

Hundreds of companies published software for these machines, giving rise to the "IBM compatible" PC. Other companies started to manufacture computers using IBM's architecture and design and Microsoft's MS-DOS operating system.

If the appearance of the mainframe was the first meteoric strike kindling the dawn of the digital age, the emergence of the PC was the first major disruption for the newly formed IT departments. It was so disruptive that many struggled to fully embrace it at first—even IBM itself, as I experienced when I started working for them.

It was the early 1990s. I arrived at IBM's impressive Bedfont Lakes facility, west of London, for my first day as a graduate hire. However, when I pulled my portable PC from my briefcase (it took up most of my briefcase!), I was promptly informed that I wouldn't be allowed to use a PC here.

"You'll be on the mainframe," I was told, "and you'll use NOSS (National Office Support Service) for your documents, email, and pretty much everything else."

Even the inventor of the most popular personal computer hadn't yet woken up to its enormous potential!

IBM had generated so much revenue from its core mainframe and minicomputer products that it didn't pay attention to the burgeoning personal computer market. It had become too comfortable selling large systems to IT managers who were cozy in their relationship with the company and failed to adapt to a changing world. (And that, despite being one of the instigators of that change!)

Although I didn't realize it at the time, we were experiencing one of the earliest major meteor strikes for IT departments: the migration away from centralized computing to personal computing. IBM's fall from grace as the world's top IT company coincided with this.

IBM was operating with dinosaurs in the driving seat! And it paid for it.

Client/Server Architecture

The next phase of evolution moved quickly, perhaps because it combined the best of what had come before: unwieldy but central mainframes, and smaller but less powerful PCs. Microsoft and Intel took the lead here, integrating Intel's efficient PC microprocessors with Microsoft's operating systems. In 1993, Microsoft introduced their "server for the masses," the Windows NT Server 3.1.

By doing so, they created a powerful ecosystem built around servers: dedicated central computers for network systems, functioning in some ways similar to how mainframes functioned. However, instead of connecting to dummy terminals, other computers (including employees' own PCs) could tap into these servers.

This newest iteration of business computer systems made office computing cheap and efficient. The client/server architecture and associated networks made it easier for employees to share data, access software, and work collaboratively on documents. A new generation of IT professionals quickly became comfortable with this latest system of readily available computers, changing once again how business computing was done.

Internet and the Cloud

And then, on April 30, 1993, one of the most significant meteors ever hit us when the internet as we know it became available to the public via the World Wide Web. This event impacted not only IT departments, but all of us; it has thoroughly altered life for almost every person on our planet today. The first time that I dialed into the WWW on a creaky old USRobotics modem in 1994, I was hooked. The ability to search for information in a seamless, intuitive way, the opportunity to easily connect with people all over the world, and the global nature of the WWW was both captivating and exhilarating. I sat for many hours bouncing from page to page, my face illuminated by the gentle glow of my little monitor. I knew that the world would never be the same again.

By 2006, we had also entered the cloud—business units no longer needed to connect into the databases and back-office systems, built by their own internal IT professionals, housed in old-fashioned data centers, and using now-outdated client-server models. In the cloud, they had access to databases, applications, and systems developed by some of the world's leading technology companies, such as Amazon, Google, and Microsoft, at their fingertips—and they needed nothing more than a credit card to get started.

The cloud both was, and still is, an impactful and rapidly changing force that business leaders need to acknowledge—and that tech leaders need to keep up with. We will talk much more about changing business models and the cloud in later chapters.

EVEN THE SMALLEST METEORS MATTER

So much has changed, and so fast, that even my above whirlwind tour through the changes covers a lot of territory. To recap, I'd like to also provide this summary of some of the significant technological events which have driven change both in the world of IT and in the business world as a whole over the last century:

- the invention and adoption of mainframe computing (the first step in the information revolution): mainframe computers became readily available for companies and organizations
- the appearance of PCs: the arrival of personal computers that sat on people's desktops democratized computing and profoundly changed the nature of work
- the creation of client/server architecture and computer networks: the combination of centralized mainframe and server capabilities and PC flexibility unlocked another wave of innovation, giving users even more power
- the creation of the internet and opportunities to work on the cloud: the internet allowed users to access information beyond the walls of their organization, connecting humans—and human knowledge—in a way that seemed impossible before.

These are the big-ticket items, the monumental drivers of change in business equivalent to the volcanoes and meteor impacts that forced life on earth to respond over geological time. However, not all meteors are large. Even small changes in technology will force change and adaptation. Think of the changes in your own office environment following the appearance of ubiquitous Wi-Fi. Or how the evolution of computing languages suddenly made DIY website-building available to small business owners and individuals who did not have the budget to pay website designers, via platforms such as WordPress and Blogger.

Even tiny meteors will drive change. Business leaders can choose whether to be agile and forward-thinking, so they can adapt rapidly to occupy new niches and opportunities, or to lose out to competitors. (Spoiler alert: The early adopters tend to win here—not the dinosaurs.)

The true tech leaders need to be doing much more than paying attention to digital systems in their office or their marketplace. They need to be looking outward and forward, spotting those meteors and asteroids while they are still way out there, and preparing to react before the competition has even noticed them.

Some meteors may not be foreseeable at all, while others start out very, very small. That's why we must always be agile and prepared to react.

TWO

Identifying Your Dinosaurs

S O WE CAN SEE THAT MANY METEORS HAVE hit the business world in the past, and just as surely, more will come our way in the future. And we all want to avoid the fate of the original dinosaurs when their meteors hit— the ones that went extinct. We'd much rather cast our lot with the new forms that emerged, the birds, which adapted to the changed world. And with good reason.

That's easier said than done, of course, since we may not even be aware of the dinosaurs lurking in our own organizations. So our first task is to take an honest inventory of our own houses.

Looking closely, you may be surprised to find you have dinosaurs that can no longer serve any purpose in today's world, and you may be faced with retiring them. But you may also discover that you have many other dinosaurs in your ranks that can be made useful through evolution and adaptation.

IDENTIFYING DINOSAURS

In order to work with our dinosaurs we first need to be able to identify them. Here's a rudimentary field guide:

Human Dinosaurs

It may seem indelicate to refer to people as dinosaurs. The reality of driving digital transformation, however, is that people play a critical role in every aspect. From your executives to the engineers and assistants on the tech team, you will encounter people who do not buy into your vision. Not only may they fail to support it, but they may also even actively resist change. It's important to work to bring these dinosaurs on board.

LEADERSHIP

The first time I (knowingly) encountered a human dinosaur was in 2016, when I returned to the UK as the new global digital director for Marks and Spencer. At the helm of the sixty-five-thousand-person organization was a visionary, forward-thinking

CEO. He explained to me his challenge: trying to invoke change in a workforce that suffered from a culture of risk aversion and inaction. He gave me his explicit encouragement to push forward and help him drive change.

One of my first observations was their smartphone app's inefficiency. It had been developed for customers to use in-store, to check prices, find product information, check stock levels, and ultimately make purchases. But it had not exactly set their world on fire.

With a little bit of research, I discovered that the median age of the company's clientele was sixty-five, not exactly the smartphone demographic. However, tablets, with their larger font sizes, would be a viable alternative. I rapidly consulted with the engineering team, and with both the CEO and CFO on board, we announced our plan to launch this new initiative in just four weeks.

But the CMO pushed back hard. Four weeks? This dinosaur—not a digital native, and highly risk averse—wanted to work like the old times: set up a committee to explore, hire a consulting firm, do a full market audit and design review, and present the results of the study to the board before approving anything.

His resistance slowed our momentum. I managed to escalate this situation to the CEO, who supported me. Within weeks we had launched the tablet version of the app, with smashing results attesting to its success: an 80 percent increase in gross revenue from the brand's app within its first month.

And my dinosauric CMO? Well, suddenly he was on board. In fact, he presented it himself at many conferences, stressing how crucial digitalization was to the transformation "he" was driving.

Pushing forward is sometimes the best way to deal with inhibitors. And allow them to bask in the results, too, if that's what you need to do to bring them on board. The proof is in the pudding: Leaders who refuse to acknowledge the realities of innovation and change, such as the former executive teams at BlackBerry and Nokia, will be forced out of the game.

PEERS

It is not only the higher-ups who may be technological dinosaurs, resistant to change. You may find that some of your peers are reluctant to come on board as well.

It is important to try to determine the roots of their reluctance. Often, the resistance has emotional roots, even for people in the corporate world, and even for many mature workers who have lived in that world for decades. Quite often, resistance to change comes from fear of change: fear of revealing that you don't know something, fear of losing status, fear of change itself.

It's common to fear what you don't fully understand. If your dinosauric colleagues are fearful because they do not have enough information about the underlying technologies, give it to them. Also explain what your strategies for change are, and why they are necessary.

Provide them with training, even if you don't label it as such. Further, you can use your understanding that your colleagues' resistance to change is rooted in fear to remind yourself to act with empathy, rather than with frustration, when trying to bring them around.

With this in mind, you can work at countering resistance to change through both communication and education.

THE TECH TEAM

It is to be expected that the people on the leading edge of the tech team will offer the most resistance. That's only natural, for they are the ones who have the most to lose. If their cherished computing systems are taken away, and their functions moved to the cloud, will those people still have jobs? Or will their positions be replaced by machines?

You will need to reassure employees about their personal job security. Show them exactly what their new situation will be, the shape of which comes down to the specifics of your own organization's digital transformation plan. Typically, it involves some combination of re-skilling or upskilling your workers and/or outsourcing them. (Much more on outsourcing to come.)

It might seem scary to team members, but these roles must evolve. The dinosaurs of IT, our old CIOs, once solely focused on engineering and specs and hardware, must become our new IT leaders, our CDOs, with tech savvy and also a forward-looking and outward-looking vision. They must play an active role in leading the entire organization, not just minding the tech team. They must bring a vision to their team and be able to communicate that vision in a way that inspires their people to follow them.

Once you've identified these dinosaurs, you can help them see this necessary transformation as an opportunity. Once again, communication is key. You need to inspire your tech team and bring them on board, get people to rally around this beautiful new thing you are building rather than trying to dazzle them with specs and performances.

Technological Dinosaurs

For most companies, the technological dinosaurs are the legacy applications and the legacy infrastructure (or hardware) which comprise those systems. Whereas you may be able to find ways to work with some of your human dinosaurs, encouraging and enabling them to adapt, hanging on to your legacy systems will only hold you back.

One big reason why: Incorporation of AI is no longer optional for any business that intends to remain relevant moving forward. Any organization that does not incorporate AI into its operational strategies will lose out to the competition. And AI algorithms require an integrated structure—a single data architecture rooted on platforms rather than individual systems.

Therefore, ditching your technological dinosaurs requires that you move your operations to the cloud. Outsourcing is the best model for large corporations to do this, and companies like Google and Accenture are generally happy to purchase your legacy systems and then outsource their services to you. (Their high level of expertise coupled with their overall efficiencies of scale allow them to provide those services at reduced cost to your organizations while still generating a profit themselves.) We'll explore your options in detail in Part III: Transform.

Operational Dinosaurs

The human dinosaurs and the technological dinosaurs are common to every organization, and overall are easy to identify. However, some dinosaurs are much rarer—with more restricted habitat or range. Identifying them might require a bit more investigation and stalking around and peering through the bushes to find them. Some ideas of where to look:

OLD OPERATING MODELS

Our IT operating model must also evolve, because the old models of systems, carefully guarded and cared for by IT teams, no longer work. This operational model is a path to extinction; fortunately, a promising future lies in platforms based in the cloud. In Part III of this book, we'll explore every corner of the cloud and discover why it is not just our friend, but our best friend, enabling us to use new technologies like AI and even newer technologies yet to come.

PAST PROCESSES

You've probably heard this defense of old processes in your organization before: "Well, this is how things are always done." We need to stop looking at our business world this way. Times they are a-changing, and change necessitates that we invent, and invest in, different methods for yielding results or maximizing efficiencies.

Using the systems-based waterfall approach for projects, where each step is reliant on every step that came before, is outdated and risky. Processes where funding is allocated to individual short-lived projects, rather than to long-lived platforms, are outdated, risky, and low yielding in terms of ROI. Internal processes where the IT department act as gatekeepers to innovation rather than as partners to business are dinosaurs.

Taking a fresh approach at defining our best processes requires us to first identify the antiquated process of the past—the places where "business as usual" is actually holding operations back. With that new perspective, we can then think outside of the box to come up with new processes and ideas.

STALE STRUCTURE

As the world changes around you, you will one day find that you cannot keep doing what you used to do. Or perhaps you *can* keep doing what you used to do, provided that you find a different manner for going about it. Walmart found this out the hard way.

Walmart is the world's largest company by revenue. It was founded in 1962 on a business model of driving sales by cutting out the middleman, thereby providing retail goods to consumers at lower prices. This strategy provided immediate success: By 1981, Walmart was the largest company in the USA, and it proceeded to expand globally.

Then a meteor hit—the internet. It was a meteor that Walmart tried very hard to ignore.

The appearance of the internet hit hard and left Walmart struggling, because their brick-and-mortar stores started losing sales to the new kid on the online block, Amazon. Although Walmart played half-heartedly with online sales, it was unable to keep up. In 2015 Amazon overtook Walmart as the most valuable retailer in the USA by market capitalization.

Walmart finally realized that it needed to leap feet-first into the world of online sales. In 2016, Walmart purchased online retail start-up Jet.com, acquiring it (and its founder, Marc Lore) in a deal that cost it $3 billion in cash and $300 million in Walmart stock.

Walmart, partnered with Jet.com, was back in the game. It had found a new way to do its old thing. (And just in time! Little did they know it, but the meteor called COVID-19 was about to hit.)

OUTSIDE RESISTANCE

In some cases, the dinosaur holding you back may not be within your organization at all: It may come from outside.

The global ride-sharing company Uber is an example of just this. Uber appeared and expanded rapidly into a scene where, previously, the only way of contracting a ride was by booking a taxi or a limo—a heavily regulated field.

The company met with substantial resistance from taxi drivers, and, around the world, Uber has had to lobby governments to devise legal ways for it to operate. Likewise, the status of Uber drivers, whether they are employees or independent contractors, is also being debated, and in some jurisdictions Uber has had to lobby to revise legislation around these definitions.

The first step toward solving a problem is admitting that you have a problem. Our world is changing incredibly rapidly. Some of that change is relatively even-paced and foreseeable, like the ever-growing capabilities of AI. Other changes slam into us without any warning or precedent, like the COVID pandemic.

THREE

Getting Ready *for* *the* Next Meteor

NCOVERING YOUR DINOSAURS IS ONLY THE first step in preparing for the next, inevitable, meteor— big or small. You must also be able to move forward fast when it does hit. In this chapter, we'll look at how some companies did a fantastic job of this when a *huge* threat confronted organizations around the world; let's go back to 2020 and the arrival of the novel coronavirus.

The near-immediate response was, like the COVID-19 crisis itself, huge. Over the first two weeks of March 2020, an astonishing burst of digital transformation occurred, perhaps even more than we had witnessed over the entire previous decade. Large numbers of the world's workforce abruptly started to work from home, and online meeting technologies like Zoom took off. Executives were forced to wake up to a new reality that threw precedents out the window.

The pandemic upended entire industries. It dethroned sector giants, and it redefined the status quo of best practices and profit-generating business models. The effects of the COVID-19 pandemic reverberated around the globe for years. For some businesses, the pandemic was the ultimate catalyst for digital transformation. For others, however, it was the nail in the coffin. We can learn a lot by studying how different organizations responded to it.

There were two new basic truths to grapple with:

1. the rules of the game had changed, and
2. no one was going to tell you what those new rules were— you were a part of making them.

Some industries, such as brick-and-mortar retailers and restaurants, were faced with a sudden and unforeseen drop in demand for their products, or with obstacles to their delivery. Others, such as Zoom and Netflix and the telecom providers, experienced the opposite: They had to learn how to cope with unprecedented demand for their products or services. Yet others needed to adapt and take advantage of new niches that suddenly appeared—or at least be prepared to rapidly move in a different direction within the niche they already occupied, such as restaurants and food delivery services.

The companies that survived, or even thrived, were those that immediately comprehended that the only way forward was through rapid adaptation.

A METEORIC RESPONSE

We can learn a lot about how to be ready for the next meteor by looking at successful responses to previous strikes. In this section, we'll be exploring some enlightening case histories from a range of companies and other organizations that were able to respond quickly and positively to the meteor called COVID.

The Biotech Race Was On: Moderna

With the start of the pandemic, one of humanity's top priorities suddenly became the development of a vaccine. Here was a newly created opportunity or niche for which there was a clear global demand: Whoever could both develop an effective vaccine *and* be able to deliver it at scale would reap great financial rewards.

Prior to the COVID pandemic, the idea that anyone could develop and distribute a vaccine in a matter of months would have been called the stuff of science fiction. But now, the race was on. Numerous players entered, employing different technologies.

Moderna, launched in 2010 as a portfolio company within biotechnology start-up Flagship Pioneering, was one of the lead players. Moderna was uniquely positioned to meet the challenge, as it was created in pursuit of developing messenger RNA technologies. Even though Stéphane Bancel, CEO of the company, humbly categorized Moderna as "a tech company that happens to do biology," it was Moderna's years of investment into research and development, along with its data-centric operating model, which allowed it to develop its mRNA coronavirus vaccine so quickly.

The company was already prepared: As soon as it had access to this virus's genetic code, it would be able to develop a vaccine for it. In fact, it took Moderna only two days to design the vaccine, back in January 2020. That was long before many people had even heard of the new coronavirus!

Moderna's vaccine was given emergency approval by the FDA on December 17, 2020—less than one year into the coronavirus pandemic.

Yes, Moderna was able to respond exceptionally rapidly to the unexpected meteor called COVID. However, Moderna's success was enabled by a decade of prior investment into technology, and by its philosophy of existing always on the forefront of the digital world. This company was ready to jump.

Giving Consultants Superpowers Through Tech: PwC

Legacy companies, too, can be "ready to jump." PwC, for example, the world's second-largest professional services network, is the product of the 1998 merger between two accounting firms founded in 1849 and 1854: Coopers & Lybrand and Price Waterhouse (PwC). That hardly sounds like an organization that's set up to turn on a dime.

Perhaps, but PwC quickly implemented Google's Workspace, a cloud-based suite of online collaboration. This allowed the entire company to avoid any major disruptions by continuing to work remotely. As a bonus, it also led to some efficiency gains and helped the organization to deepen relationships with its customers.

PwC personnel were now able to have meetings at short notice by using Google Meet. Over 50 percent of their meetings with clients became spontaneous and ad hoc. Meetings also became shorter as attendees were able to get straight down to business. This is a great example of the pandemic forcing positive change.

There is always the option to be agile, responsive, and adaptive. PwC chose to behave like those dinosaurs that evolved into birds, proving that legacy businesses are not necessarily doomed to go the way of the dinosaurs, even when confronted with unprecedented meteoric change.

In contrast, many other companies struggled to adapt. In fact, I heard reports of companies that were not able to have board meetings at all, because they had no policies in place to even recognize virtual meetings!

Unleashing the Power of Tech for the Next Generation: Italy's Ministry of Education

The field of education was one of the areas most affected by the pandemic, and it's no mystery why: Until recently, educational institutions functioned predominantly through in-person, classroom-based, or field-based instruction. No wonder an estimated 90 percent of learners worldwide had their education disrupted by the pandemic crisis.

Italy was one of the earliest countries to be hit hard by COVID, and it experienced some of the longest and harshest lockdowns. As soon as the pandemic hit, the Italian government made a rapid turn to the power of digital to connect every school in Italy to online learning.

This was no small feat. Physically, the schools typically did not run up-to-date hardware or software. In fact, many Italian schools did not implement technological resources much at all. Additionally, the majority of their employees were not comfortable with the use of technology: Teachers still mainly used classic models of face-to-face teaching, relying very little on computer-based aids or online resources.

This system of classroom-based, face-to-face learning worked fine in Italy. There was really no need to embrace remote learning.

That is, until the sudden March 2020 school closures and lock-downs. In that moment, Italy's government stepped up and delivered. In less than one month, they launched an integrated online learning platform that might otherwise have taken years to complete. A dinosaur no more.

To achieve this, the Ministry of Education's first step was to set up a new Task Force for Educational Emergency. It made a public call for support from industry. Several telecom companies and Google responded.

The task force selected Google's G Suite for Education as a platform. In collaboration with the task force, Google developed new online distance-learning programs rapidly and from scratch. Funds were quickly allocated by the government toward purchasing devices for teachers to be able to access these new platforms.

The Ministry also recognized that the teachers themselves required support. They needed instruction on how to use these new tools. Programs to both fast-track enrollments and to support teachers were rapidly developed, and a call center was activated to help teachers who needed one-on-one support.

Through its rapid and integrated response to the unforeseen pandemic, Italy—once a dinosaur with its antiquated educational system—chose to respond and evolve, flying like a bird into the twenty-first century.

Feeding the Masses: DoorDash

Some adaptations don't require such a huge shift; organizations can often find strategies to operate in a new way within a niche they already occupy. This type of digital adaptation worked well for restaurants that were able to continue with in-house dining during the pandemic. They quickly embraced the benefits of technology such as switching to QR codes for menus, allowing employees and clients to maintain physical distancing by avoiding contact between servers and diners. Even with most pandemic restrictions dropped, many restaurants have chosen to stick with this new approach, which lowers costs by allowing them to reduce staff and enabling easier daily menu updates.

The pandemic opened up another niche in the restaurant industry as well—that of online food delivery services.

Online food delivery businesses had existed prior to the pandemic. However, it was still a fairly new industry, experiencing only moderate growth through 2018 and 2019. Demand for the service spiked abruptly in early 2020, though, as lockdowns and travel restrictions were suddenly implemented in many parts of the world.

Those who were ready to rapidly change how they did business and occupy this new niche would have the most to gain.

DoorDash is perhaps the company that was able to make the most gains during the pandemic, likely because it had already invested heavily into new ways of operating, including new tech and cloud-based systems, for no other reason than to remain on the cutting edge, and ready for...whatever. So, when the global business environment abruptly did an about-face, it was prepared and ready to go.

And go they did. In 2018, DoorDash controlled around 15 percent of the US food delivery market share. By early 2021—partly through their acquisition of competitor Postmates, but largely through expansion and growth—DoorDash was in control of well over 50 percent of the US food delivery market.

Creating Endless Aisles in Retail: IKEA

Many retail companies were not as prepared for the challenges presented by the lockdowns and other restrictions that prevented potential customers from visiting their outlets and accessing their products. Retail CEOs became painfully aware of this new truth: They had to quickly shape up and mark their presence in the e-commerce market or go out of business. This meant completely reassessing their business practices.

Not all chose to tackle and transcend their digital shortcomings. Many mom-and-pop shops went underwater, as did some of the "bigger fish" such as Neiman Marcus and JCPenney. In sharp contrast, however, the Swedish multinational furniture conglomerate IKEA did not waste any time in saving itself.

IKEA quickly closed 433 of its enormous stores, repurposing them as fulfillment centers for online orders. In only a week, through CDO Barbara Coppola's leadership, it successfully completed the migration of thirteen regional IKEA websites to the cloud—centralizing, consolidating, and integrating all these systems.

By using AI, tech, and data analysis to shape its customers' digital retail experience, specific to their regions and offering personalized recommendations to online shoppers, IKEA adapted by rapidly and completely reshaping its customers' online retail experience—continuing business as usual, pandemic or not.

On the Fly Telecommunications: Verizon

Whereas many industries faced a drop in demand for their products with the start of the pandemic, the problem was the reverse in the telecommunications field—usage of internet bandwidth surged. Companies like Verizon, in an industry where service continuity is paramount, faced the very real possibility of not being able to provide that continuity.

Verizon's strategy for survival was immediate and hands-on. It set out to create contactless service as much as possible, by deploying an app-based prebooking system for in-person store visits. In addition, it developed software that allowed technicians to assist virtually with installations and repairs.

Verizon extended the premise of contactless service to digital contract signing, payment, remote co-browsing, and ID verification, solidifying its pandemic-proof strategy. Overall, it condensed and completed what normally would have been years of digital development in just a few weeks.

In the midst of the chaos, Verizon CIO Shankar Arumugavelu explained to *Forbes* in May of 2020 how unique the situation was: "Everyone had a business recovery or business continuity plan, but these were all for more contained scenarios. We're writing the playbook on the fly."

Verizon's efforts to survive the pandemic were successful. While many organizations took a financial hit as a result of the pandemic, Verizon actually increased its adjusted earnings by 5 percent in the first quarter of 2020.

STRATEGIES FOR PREPARING FOR METEORIC CHANGE

I have presented the case studies above because they illuminate real-life positive responses to catastrophic change. These studies also illustrate many of the points that I will be discussing throughout this book: strategies that you, as tech leaders, should not only be aware of but embrace, such as:

1. Invest in Technology—Even When You Don't Know Where It May Take You.

This is how Moderna and DoorDash not only survived the pandemic but outran their competitors. Each had invested heavily in tech for years. In Moderna's case, it pioneered with innovative research strategies, combining the most recent and advanced scientific developments with AI. Similarly, DoorDash had forged ahead by acquiring and adopting the most advanced AI-based and cloud-based systems for use in its day-to-day business.

Because these companies were already on the cutting edge, and ready to go, they did not have to play catch-up when unforeseen changes hit.

2. Embrace Technology

Investing in technology even when you don't know where it is going to take you is only the first step, though. You must also embrace technology once you *do* know you need it. If it is becoming hard to do business how you have always done it, don't hesitate to find other ways to do the same business. Use technology to your advantage.

That's what the restaurant industry did by adopting QR codes at dining tables. That's how IKEA completely revamped its retail process in mere weeks. That's how Verizon found new online strategies for conducting its same business, but in different ways. They could not function as they had always done, so they adopted new technologies to change it up and find success.

3. React Quickly When Sideswiped

Whether you are already positioned on the cutting edge, or you are a bit farther back in the field, you must be able to move with speed and agility. That's what both PwC and the Italian government did.

Both had been a bit slack in the past. Neither was already perched there at the front of their fields, like Moderna and DoorDash, ready for a disaster. But when catastrophic change struck, they did not dillydally. Each moved swiftly and decisively, with a vision. In both cases, they achieved within weeks what organizations would normally budget months or even years to do. It's amazing what we can do when we are truly motivated.

4. Identify New Niches, New Opportunities

When unforeseen change happens, some businesses will fold, laying open their niches for new takers. And new niches will appear as the environment changes as well—niches that did not even exist before.

The COVID pandemic spawned a demand for a new online version of old businesses. When the market for in-house dining suddenly waned, the niche for food delivery was suddenly wide open. Many restaurants, rather than closing down completely, moved to various versions of takeout and/or delivery service. And other companies such as DoorDash moved in quickly as the food delivery market opened up.

Food delivery is but one example: Many other new niche internet-based businesses opened during the pandemic, such as at-home fitness, box subscription services (e.g., wine, beauty products), and virtual online classes and workshops.

LOOK TO THE FUTURE

We may not know exactly *when* the next meteor will hit, or what it will look like. The next big hit to our systems could be climate catastrophe, or nuclear strife, or supply chain challenges (whether caused by humans or natural disaster), or yet another pandemic. Or the next meteor may be small—yet still significant and invoking change—such as the evolution and adoption of a new technology.

What we do know: There are always more meteors coming.

So we expect the unexpected. We know that meteoric changes are ahead. Even if we don't know exactly the nature of those meteors, we can still prepare for them, and be ready to move, much like the organizations covered in this chapter.

And we *prepare* for the unexpected. Cultivate a personal philosophy to live and work by and a company culture that looks ahead, committed to being agile. Be prepared to act quickly, and also to reassess and abruptly change direction if the situation so requires.

Am I saying you need to be a visionary? Not exactly. I am saying you need to look to the future, but to me, being a "visionary" is not so much about being born with some nebulous genetic gift, but about the concrete things you can do to help prepare for the future, or even foresee it. At the same time, you must accept that your digital transformation will never be complete. It will always be a work in progress. Let's get started.

PART II

Evolve

FOUR

Defining *a* New Operating Model

THE TRANSFORMATIONS WE EXPLORED IN THE last chapter are inspiring; we all want our businesses to be able to evolve like that. I realize that "transformation" and "evolve" are weighty words, but I believe they are the right ones to describe the scope of change before us. The change goes deep—it requires questioning the long-held, though costly and outdated, business models we've relied on in the past, because they simply don't work anymore. If we want to succeed, we desperately need to define a new operating model.

Still, I know it's hard to let go of the way things used to be, especially if, once upon a time, that seemed to work splendidly. But if your organization genuinely seeks to adapt to this rapidly changing digital world, your change needs to be systemic. It's not just a matter of upgrading your server or creating some new software. Evolving will require a complete rethinking and reconfiguring of how your IT department works—and even more so, how IT interacts with the rest of your organization.

However, most companies today are not doing this! Instead, they are throwing their IT budget toward keeping their old systems alive, just patching and upgrading. Seventy percent of IT dollars are spent on unwieldy legacy computer systems. These systems are cumbersome, they are notoriously incompatible with other systems, and they rely on in-house tech. And yet, they keep throwing money into this "dungeon." Why? Let me offer an analogy—the story of feeding the dragon:

Once upon a time, there was a young monarch who was out expanding her queendom. She had a dragon that she soared around on, and she was doing a fantastic job—flying her dragon everywhere, setting fire to her adversaries, and conquering other kingdoms and queendoms hither and thither.

Eventually she vanquished her enemies. She attained the throne and became queen of all the territories.

Now peace reigned, and all was good. The queen didn't *need* her dragon anymore. However, that dragon had helped her become queen, and she still had warm feelings for it. She appreciated what it had done for her. So, she kept the dragon alive down in the dungeon, treating it well, and paying keepers to toss it the occasional sheep.

So, the dragon lived on. The dragon-keepers cared for it devotedly, and the queen was happy to pay them for their duties. Perhaps she would require the dragon's services again someday.

But over time, the queen stopped visiting her dragon. Eventually, the once-close relationship between the queen and her dragon faded. In fact, it had been relegated to others: the lowly dragon-keepers whom she paid to take care of it.

Time passed, and life in the queendom changed. The dragon now served no purpose at all. Its services were not required—and besides, there were new types of weapons available and new types of technology. Even if war *did* break out again, the queen wouldn't need her dragon. She had much more advanced military options available.

But the dragon-keepers down below saw things differently. They didn't care what was happening out in the queendom. Their job was simply to feed the dragon. In fact, their very livings depended upon keeping that dragon alive.

And on top, they had come to care for the dragon too. They didn't want to get rid of it—they wanted to protect it.

ARE YOU FEEDING THE DRAGON?

This probably sounds very familiar, and you're hardly alone; that 70 percent of technology spending we talked about earlier is essentially spent "feeding the dragon." And, of course, that dragon is your business operating system: both your hardware (mainframe, servers, what have you) and the software, all of which the dragon-keepers on your payroll lovingly maintain and feed.

We now find ourselves in a bizarre world where board members and marketing teams are practically screaming out to the IT team for new functionality and new capabilities. But the IT people are shaking their heads, saying, "No, can't do that. But we need more budget, more money, so we can keep feeding the dragon, so we can keep our beloved systems alive."

Well, what's the solution then? Upgrade to a new, more evolved dragon? Acquire a dragon that is sleeker and faster and comes with enhanced functionality?

Or accept that the days of keeping a dragon down in the dungeon are over? That dragons no longer serve a purpose? This is where we are in the digital world today. What is needed is a complete structural overhaul and adoption of an entirely different operating model.

MOVING FROM SYSTEMS TO PLATFORMS

To make the necessary shift, we need to change both how the IT department *works* and how it *interacts* with every other level of the company. This requires a complete structural revisioning and rebuilding: from an operating model founded on *systems* to an operating model based on *platforms and products*.

What Is a System, and What Is a Platform?

The distinction between systems and platforms is important—and it can be confusing—so we're going to spend most of this chapter exploring those differences. On the surface they appear similar to the point that they may seem identical, but when we look closer, we can discern a key difference. Yes, both systems and platforms consist of a complex of hardware and software. The difference between them is the vision of how that complex interacts with the organization as a whole.

Systems are point solutions, designed against a rigid set of requirements or specifications at a specific point in time. They may indeed provide a solution to that one problem they were designed for. However, they are not designed with the vision of being applicable across all departments in an organization over a long period of time. This means that systems come with both risks and costs, such as duplication between departments, incompatibility issues, and rapid obsolescence.

My dragon tale presents the example of a *systems* approach. The dragon served a purpose. It made some serious progress and achieved great success for its owner. However, the dragon really served only one purpose, for one specific period of time. Then it timed out.

Platforms, on the other hand, are long-standing solutions with long-living capabilities. They are not created to meet a particular initiative. Rather, they are designed to meet long-term generic needs across the entire enterprise, and to support all the departments, typically for a period of ten years or more.

An operating model based on *platforms and products* is one that serves multiple purposes, across multiple departments in your organization, and that is designed for the long term. Adherence to the platforms and products operating model is one key to the enduring success of companies like Apple.

The Problem with Systems

Systems are designed against a very rigid set of requirements and specifications. Your IT department receives the requirements, and they configure the infrastructure, define the database, build the application logic, and tie it all together. That's not a bad thing—in fact, it's a great way to solve one problem, for now.

But then the system gets buried in a mire of more problems, then more solutions, requiring more systems to be built, one on top of the other. This often leads to duplicate systems, each developed in-house (at great expense, I might add—both to cre-ate them, and then to continue feeding them).

For example, I worked for one major telecommunications company that had (literally!) thousands of different systems in place. They had systems designed for specific campaigns or plans, and others to handle tasks like payroll or inventory. These systems duplicated similar (but incompatible) systems that had been built by other departments. And they had systems built on top of other systems, creating a fragile ecosystem and significant cybersecurity risk. Some of their systems supported only a hand-ful of customers—but together, they consumed millions of dol-lars to operate!

One way to think of the systems approach is that it is like the game of Jenga—that tower made of layers of wooden blocks. Players must remove one block from below and replace it on top of the tower. At some point, the base of the tower becomes too weak, and the whole thing topples over.

Similarly, when you design your IT department as a system, you end up with too many parts all dependent upon what is down there at the bottom. You cannot upgrade or change with-out knocking the whole thing down and starting over. You are constrained by the antiquated capabilities of the base of your system. Eventually that base becomes too weak, and your choices are either to invest in endless patches to keep the thing standing, or to finally allow it all to topple over and restart from scratch.

Systems *can* be very good for delivering what the business end of the company asked for—but only at that one moment in time. The problem is that everything is fixed. As soon as things evolve, and there needs to be a change, problems appear, because the system has been designed from the ground up.

The long-term solution is to abolish that entire systems-based structure and move toward a structure based upon platforms and products. Let's look at how platforms work, and then at what I mean when I use the word "products." Within the context of "platforms and products," the idea of "product" may not be quite what you think.

How Platforms Work

Like a system, a platform is an integrated network of hardware or software—usually a combination of both. However, as shown in this diagram below, I visualize platforms as horizontal, just like a physical platform. Unlike that unstable, vertical Jenga configuration, where each piece is dependent upon the point solution immediately below it, a platform is broad, and it independently supports each of the pieces above.

In contrast to systems, platforms are not designed to meet a particular initiative. Rather, they are long-standing, designed to meet long-term and generic needs of the enterprise, typically over a period of ten years or more.

	Product 1	Product 2	Product 3
Platform A			
Platform B			
Platform C			

Let's go back to Apple for an example of a platform— self-service.

The self-service platform is consistent throughout all of Apple's departments. If you are an Apple customer, you can log into your account and do whatever you need to do to review your past orders, or change your address, or track an existing order. It doesn't matter whether you want to purchase a new computer, or buy an AirTag, or stream Apple TV. That self-service *experience* is both familiar and efficient for all users, over a long period of time, over all the world.

Another example of a platform at Apple is payments. Likewise, it is consistent throughout all departments, no matter where you are in the world.

There actually is a "law" at Apple that requires that all experiences must be done via Apple's existing platforms! For example, if you were in charge of a new business unit at Apple, and you would like to implement a service like order tracking—by Apple law, you would not be allowed to build that tracking system yourself. You must use the existing platform.

And, of course, this leads to great efficiencies, both in time and in cost. When Apple wants to develop something new, they simply build a technical product on top of their existing platforms. Whereas with that systems-based telecom I mentioned above—well, every time anyone from business asks for something, IT goes and builds them a new system from the ground up. Last I checked, that telecom was running 4,600 different systems!

And the people running those systems, those dragon-keepers, stand protectively on guard. These monolithic things were lovingly created over many years. It's the natural instinct of the people maintaining them to resist change, to protect them. "Don't change anything! Or everything will stop working!"

Products—Supported by Platforms

In the business world, we often think of "products" as part of "products and services"—that which we lease or sell to generate income, such as cameras or corn or software or transportation services or technical expertise.

However, in the context of "platforms and products," the word "product" refers not to a product for sale, but rather to a digital tool used internally by the organization to improve efficiency or capability. Amazon calls these "technical products." That is terminology that makes sense to me, distinguishing these in-house technical products from products for sale, so I will refer to them as such here.

Here is an example of a technical product. When I first joined Apple in 2011, one of the problems the company was facing was finding a way to help customers find the right cable to connect their devices. People would walk into an Apple store, where friendly Apple employees were eager to try to help them by inquiring, "What cable do you need?"

But then the customers would reply, "Well, I'm not sure." Of course, they didn't know which cable they needed—that's why they were asking for help! They only knew which devices they were trying to connect.

So, Apple's business unit asked my team to build a technical product—a piece of software—so that a customer could simply specify their need: "I am trying to connect A to B; which cable do I need?" The software would spit out the answer. The customer could then go on to make their purchase without delay or hassle or error and go home with their purchase.

This technical product was nothing for sale—it was simply a piece of software with a very specific purpose, and which worked on top of Apple's existing platforms.

Another example of a technical product I collaborated on at Apple was one that tracks delivery times. The fulfillment department wanted to introduce a new service to allow customers to get their Apple products delivered within hours instead of days. The technical product we created provided them with a fast and easy way to do that—customers in some cities can now get their products delivered within two hours!

Technical products work on top of existing platforms. Whereas platforms are designed to be long-standing and broadly based, technical products solve very specific problems by enhancing user experiences.

WHY THE FUTURE IS PLATFORMS AND PRODUCTS

So far, we've talked about the main advantages to working with platforms: There are cost efficiencies. The incompatibility issues are eliminated. User experiences are standardized, so learning curves are much reduced. And a platform base allows for easy upgrades, or even outright changes or replacement. That may well be enough to convince you of the value of a company IT structure based upon platforms rather than systems.

But there are many further benefits to working with platforms, chief among them the reality that everything changes, and it can change quickly—and well-thought-out platforms and products allow us to be ready for whatever comes next. Having long-lasting platform capabilities that are designed as the foundation for change allows you to evolve rapidly—and cost-effectively. Fixed, monolithic systems are resistant to change—and a threat to your organization's ability to adapt.

Let's go back to that Steve Jobs speech, introducing the first iPhone. He is talking about his new revolutionary user interface, the touch screen. He shows a slide with the competition's phones: the Motorola Q, the BlackBerry, the Palm Treo, and the Nokia E62. Each of these sports a miniature QWERTY keyboard with tiny plastic buttons. "What's wrong with their user interfaces?" Jobs asks—then, of course, answers his question for us.

"They all have these control buttons that are fixed in plastic and are the same for every application!" However, he explains, that arrangement of buttons may not be ideal for every application. "What happens if you think of a great idea six months from now? You can't run around and add a button to these things. They're already shipped!"

Apple's solution: a new touch-screen platform that allowed for infinite variations in where buttons can be placed, how many are needed, their size, their shape, their color. That forward-looking platform has proven, nearly two decades and a billion and a half iPhones later, to withstand change and be adaptable to new software and new products which were inconceivable at that time—even to Jobs himself.

Apple is the master of working platforms and products rather than systems. Their adherence to the platform structure—whether for in-house company initiatives such as their self-service and payments platforms, or the flexible user interface on a new product for sale—is a huge contributor to their ongoing success.

Our Job Is to Create on Top *of the Platform*

How much better does a platform-based approach work? Let's see. Here's a story of two smartphone sellers. One—already mentioned—is deeply embedded in the systems approach. Their marketing team wanted to introduce a new type of promotion.

So, they took their concept to the IT department. But the IT department told them, "Oh, our system doesn't do that. We would have to change the system." The IT department demanded a set of requirements from marketing, then said it would need eighteen months to go away and think about how to change the system—oh, and by the way, changing the system was going to be very expensive and highly risky.

Ultimately, the company gave up on the initiative: They decided to use human-ware and handle promotion through call centers instead of through technology. They were caught in a familiar cycle: Instead of the IT function *responding* to a marketing need, the IT department was *driving* what the company was actually able to do.

The outcome of the last initiative I was part of while at Apple was quite a contrast, even though we started with a similar proposition—delivering the iPhone upgrade program. The goal of the program was that a customer could come in, buy an iPhone online, then every two years they could upgrade that phone—with all our software managing the entire process.

We were able to complete that initiative in just nine months, for a fraction of the cost that telecommunications companies incur for delivering fewer satisfying experiences, because of the platforms and product paradigm. Today, 70 percent of the profitability at Apple comes from that one initiative.

Bizarrely, it is actually more expensive to purchase an iPhone this way than through a telecom provider. However, since we worked backward from the needs of the customer and designed such a beautiful experience, it was a great success!

THE ROLE OF THE IT PEOPLE HAS CHANGED

In light of the profound paradigm shift I'm describing, it's not surprising that the role of the IT leader has changed drastically, and we've discussed some of those changes already. At the same time, we also need to look at role changes for the entire IT team, and how the role of each employee has also been transformed.

Back in the nineties, and even in the early aughts, an IT person was essentially an engineer, focused on systems. They were not a generalist, but a deep expert in something: some component of the system. If you asked that IT person what their job was, they would tell you, "I am in charge of System X." And they were a wizard; they knew every aspect of System X.

Back in the nineties, that was great.

Today, as we've seen, most of the tech that the big corporations are running is still based on these very deep systems, initiated decades ago, and which their tech people are very wedded to. The problem with this structure is that, instead of IT function responding to the enterprise's commercial needs, the IT team (those dragon-keepers) act as protectors of their system.

Companies like Apple, on the other hand, don't have systems at all—they use platforms. And they don't necessarily create it from scratch, in-house. This may surprise you, but it's true; even tech companies like Apple and Google buy much of the tech that they use. For example, a few years back, Google purchased SAP for their Enterprise Resource Planning (ERP) rather than try to develop equivalent technology in-house.

So...if your IT department is no longer there to manage systems or to develop software...well, what are they actually supposed to do?

First of all, the modern IT leader must approach the future looking through the eyes of their end users or their customers. Remember my example of how we developed the iPhone upgrade program? We worked backward from the needs of the customer, then developed the technology from there—on top of the existing platform. This is the opposite of how things have been traditionally done, where IT dictated how you can (or cannot) move forward.

The result of this change in direction is radical changes in the roles of your IT personnel. Now, instead of IT people managing individual systems which they develop deep expertise in (and defend!), they move over to one of two general roles. Some will be assigned to managing the platforms. Others will be tasked with rapidly delivering the digital experiences (in other words, the technical products) that are built on top of the platforms.

The duties of the new IT people are significantly different from those of the former system engineers. Making this move may require skills upgrades and training. Or it may mean a complete turnover. You may find yourself outsourcing your people to your tech partner. Change is hard, and this one will challenge both business leaders and IT leaders as they confront the new reality. I'll have some examples in the following chapters of how several large corporations have already dealt with this challenge.

INTEGRATING IT INTO YOUR ORGANIZATION

I cannot emphasize enough the massive business climate change we are facing, or how complete our response must be, so I'll say it again: *The entire operating model for integrating IT into our organizations has changed.*

First of all, the relationship between business leaders—the CEO and other executives—and the tech leader needs to change. The business leaders need to leave behind the custom of issuing orders or developing requirements to present to their IT leaders, who then mull over the requests for months or years (or, instead, try to adapt them and shoehorn them into their existing systems).

Rather, IT leaders must start working much more closely than they traditionally have *with their business stakeholders*. Instead of looking at things through the lens of their "systems," modern IT leaders must look at things through the eyes of the end users or the customers. The IT leaders need to be a part of defining and shaping the things that need to be built, rather than simply taking orders.

Ditching the systems and implementing a platforms and products structure then allows the creation of the customer experiences. The new IT team, no longer relegated to defending their systems and feeding their dragons (and perhaps with a bit of new training under their belt), focuses on developing the products that build consumer experiences, which translate directly to profitability.

Easier said than done—I know. In the next chapter, we will look more deeply at how to begin to adapt the company structure and relationship in order to accommodate this new and different operating model.

FIVE

Evolving *a* New
IT Organization

N TODAY'S WORLD, ANY LEADER SHOULD BE
able to work collaboratively with their IT team, striving
together toward achieving their shared goals.

And yet, I recently traveled to Texas to work with the leader
of a large Fortune 50 company who was frustrated because this
just wasn't the way it was working for her. In her company, she
felt shackled, unable to implement meaningful change. Every
time she wanted to create something—say, a tracking software—
the IT department would ask her to write everything out. She
would spend days (or weeks) writing requirements, then wait

for months while they perused the documents, only for them to eventually tell her that their system actually won't allow what she wants.

Rather than a working partnership, the whole dynamic was functioning more like some sort of external relationship—as if the enterprise leader was ordering products and awaiting confirmation from an external supplier rather than working cooperatively.

By now we've established that this process is backward—and completely wrong—but it still happens. Here's how to do it differently:

When I was at Apple, one of my tasks was to work with the product manager, Colleen. She was in the retail organization, which meant that her position required lots of support from IT.

Here was our process: If Colleen wanted to create something, she and I would physically sit down together in a room and brainstorm. We would talk about what she wanted, perhaps whiteboarding it, until we were sure that I fully understood exactly what her vision was and what her needs were.

Then I would ask my engineers to go out and build a prototype of the software she needed.

Once we had that prototype up and running, Colleen and I would get together again and look at what needed tweaking. And just like that, we were done! Within a matter of weeks, we would have a completed and tested technical product, ready to put into use.

That approach horrifies the majority of tech leaders!

"What do you mean?" they ask me. "You're just playing around, creating software?"

Yes, I exclaim, that's exactly what we're doing!

That's how the modern world needs to work. The priority must not be on spending weeks or months on documentation, but on creating working software.

Getting there requires a healthier relationship between our business leaders and our tech leaders, a relationship that will need to be supported by new and innovative ways of structuring our organizations. And that means major changes in how our IT departments interact with other areas of our business organizations.

IT'S NO LONGER ABOUT THE COMPUTING

Quite possibly, this sort of organizational change might mean looking at different skill sets for our tech people too.

Back in the days of systems, the tech people each had their clearly defined specialties: hardware, databases, coding, network, etc. They took pride in being specialists within those individual domains.

In the new world, we can do all the tasks our tech people and engineers used to do—and do them better and faster—in the cloud. We don't need all those specialists (at least not within our own company).

What's more, by working in the cloud, we aren't tied to our systems and in-house specialists anymore. If what we are doing in the cloud isn't working for us, we don't need to keep feeding that poor aging dragon. We literally can ditch our system and move to something else.

This wasn't (and isn't) always so. The problem that my collaborator in Texas was facing was that each "systems island" was governed by an IT specialist. If she wanted them to build something for her, she had to paddle over to the island, explain what she wanted, and hope they had the capability to produce it with the resources they had on their island.

Back in the nineties, this way of doing things made sense. Technology was still a bit of a secret art, and those IT specialists held the key. They had the freedom to decide what both could and should be done.

But things are very different these days! Your typical marketing leader or finance leader has a pretty good understanding of technology. They use the internet and apps themselves, and they understand the capabilities and the potential. They are absolutely not interested in paddling around to different islands. They just need stuff done.

So what of the people stranded on those islands—the tech people? What has changed, and what must change, for them?

The big difference is that their focus should no longer be on the computing. Rather, the IT department needs to be focused upon *outcomes*.

For an example, let's say you have a tech person who is responsible for customer support. In the old model, the tech person took care of the computing system behind customer support. In the modern world, however, the role of the IT person in customer support would literally be to take care of customers. To support customers. Their job is to be super focused on driving the absolutely highest level of customer satisfaction with the experiences that they build. Nobody cares what computing processes they use.

This represents a complete shift in what your IT department and your IT people need to do. It doesn't mean there's no place for tech people in our future. There is. However, their work, and the way they work, will need to be aligned to something specific: a business need, or a group of customers, or a particular outcome.

Meanwhile, the old IT role is simply a dinosaur that you need to leave behind. As a transformational leader, you may have to continue working with some of your dinosaurs, but you must let your legacy IT go extinct—continuing to invest in it will deliver diminishing returns and compromise your development.

This evolution will not be easy. It is perhaps the biggest obstacles that you, the C-suite readers of this book, will have to face. It can be done though—some high-performing organizations such as Apple and National Geographic have already made the shift. More will follow. And, quite honestly, I believe that organizations that are not willing to adapt to this new world will be left behind.

THE NEW IT TEAM

So, what will the new IT team do, then?

We will still have IT people with familiar-sounding roles: tech people who specialize in areas like fulfillment, or business customers, or databases. The difference, however, is that the tech people will no longer work from the *tech* point of view, or deal with any systems at all. Instead, they'll work in their roles from the *business* point of view. Their focus is now entirely based on *experiences* and *outcomes*: How can their organization use technology to deliver the best possible experiences for users?

Prioritize Collaboration

For example, when I worked with Apple's retail product manager, Colleen, as her tech lead, Colleen's responsibility was to deliver financial outcomes: She was responsible for P&L and working to maximize revenue. Naturally, she required a lot of support from IT in order to deliver these outcomes.

Colleen needed a direct line to the tech people who build the software she needed—someone who would outlast *my* time with her—so she'd never again find herself whiling her time away writing requirements and awaiting some sort of response. To facilitate this new line of communication, we created an entirely new IT position in her department. We called it a technical product owner.

The person we selected for this role was a software engineer and analyst, someone who already understood Apple's platforms and how they worked. Their new role was to work entirely one-on-one with Colleen, to spend all day, every day, working with Colleen, in order to thoroughly understand her business requirements and the imperatives.

This new role partly meant serving as a translator, facilitating communication between business needs and IT capabilities. However, it involved much more than merely translating; it also meant bringing insight and creativity and ideas to the table, providing feedback and ideas and vision to Colleen, and opening Colleen's view to what was actually conceivable and possible and doable.

As a team, Colleen and her technical product owner could brainstorm ideas and shape them and envision outcomes and possibilities, working together in a way that was both collaborative and dynamic—it's a whole new operating model.

To move into this new operating model, IT departments require a new structure, and the individuals on the IT team must acquire new skills and take on a new point of view. IT professionals will need to see themselves as partners to their business stakeholders, and thus be outward facing rather than simply the owners of systems.

Of course, tech professionals must still have technical expertise and use that expertise and experience to help shape organizational strategy. However, it's increasingly important that tech people are also great communicators and have a solid understanding of the business drivers that shape their organization.

Why this focus on communication and deep understanding? Once again, it's because things have changed and will continue to change. While we once operated on a "pull" model, where IT waits for business units to request things, that's no longer the case; moving forward, the best-in-class IT functions will become "push" organizations. They will work closely with business units—even embedding with business personnel, as we did by providing a technical product owner with Colleen over at Apple. In that way, embedded IT people become internal trusted advisors, enabling strategy to the business section.

Above all, the CDO or tech leader must be one of the most active members of the leadership team because of the importance of digital and cybersecurity.

Embracing Agile

The concept of working this way—prioritizing collaboration and results over paperwork and documentation—is not the one most business organizations use today. However, this more agile approach is not by any means a new concept.

You may know where I'm going with this; many of you will already be acquainted with the Agile Manifesto. If not, here's a quick overview:

Over twenty years ago (a long, long time ago in the world of tech!), way back in 2001, a group of seventeen software developers organized a meeting in Utah. They wanted to discuss the future of their industry. They emerged from their retreat with a short but impactful document that they called the Agile Manifesto, and which delineated a set of values, prioritizing:

- Individuals and interactions over processes and tools.
- Working software over comprehensive documentation.
- Customer collaboration over contract negotiation.
- Responding to change over following a plan.

The Agile Manifesto was written as guidelines specifically aimed at software development. However, these values pertain exactly to what I am talking about here with regard to business culture.

Agile has several salient characteristics. It has:

- a dynamic and flexible architecture based on platforms and products.
- fast-moving development teams that create high-quality user-oriented products, leveraging existing platforms rather than building new point systems.
- a mutual language, with definitions based upon a shared understanding of both the business needs and the capabilities of the technology.

Agile processes are exactly what we need to develop an entirely new relationship between business and IT. Otherwise, we remain dependent on the old "waterfall" approach, named as such because it is a cascade: Step A leads to Step B, which leads to Step C, and so on. Working this way, however, we find ourselves again mired in a "systems" approach, where each component of the system is based on something else prior or below or earlier. Nothing can be modified or created without undoing what has come before in the cascade—the opposite of agile!

For many (if not most) business organizations, a shift from this traditional approach to an agile approach is a critical adaptation. By embracing agile, IT leaders can get away from the relentless draw of the waterfall. They can start by putting the needs of the business ahead of any demands by IT for documentation or requirements. With that mindset front and center, they will be ready to work with business leaders.

THE POWER OF PARTNERS

So, if our IT team is now getting all touchy-feely and worrying about outcomes and communication rather than hardware and coding...well, then, who takes care of the actual computing?

Simple: Your partners do.

The way to get the best people working for you is to contract out most, if not all, of the low-level tech functions. The top tech people work for the big tech companies, not for you. So, you need to partner with them.

Your tech people can then change their focus toward experiences rather than computer systems. This is a point I have perhaps already belabored, but it's important enough to repeat it here: You no longer want to have those systems in your organizations at all. No more feeding the dragon.

Once you've transformed your company this way, your own people can focus on creating value. This does mean a very different way of working for your tech team, partnering much more with business and participating much more in developing a vision and coming up with creative solutions.

Successful Tech Partnerships: How Deutsche Bank Did It

We've looked at some examples of recent IT success stories, and you may have noticed one thing they all have in common: The key to their success was that tech partnerships were central to realizing their digital transformation plans.

National Geographic, Brit/Ki, and the Italian Ministry of Education all contracted Google. DoorDash worked with Amazon. Even the old legacy brewing company, Carlsberg, assembled a team, with Accenture planning and overseeing the whole digital transformation, Microsoft providing the cloud platform, Avanade executing the cloud migration, and Zscaler overseeing online security.

Much of the cost savings of migrating tasks to the cloud result from reducing computing infrastructure (e.g., shutting down data centers) and reducing staff. But if we are assigning the main tasks that our tech team used to do over to contractors—well, what do we do with our tech people?

Here is how one company did it:

Deutsche Bank, a global giant in the world of finance, was founded in 1870. However, by 2020, after 150 years of operation, this legacy company found itself in trouble. Despite spending billions on technology, they were encountering endless problems with functionality, along with constant issues in cybersecurity.

The business structure in Germany is somewhat different from most other countries. German companies are required to have "works councils" that give employees a voice when the business is considering any major decisions. While this system is intended to help protect workers, a side effect is that it becomes difficult for management to initiate changes in any way that might seem "agile."

Deutsche Bank was struggling. They were spending billions on tech and getting nowhere. So, my old boss at Google Cloud, Thomas Kurian, contacted them with an offer to help.

The plan was for Google to migrate all Deutsche Bank's operations to the cloud. Google would charge them significantly less than what they had previously been paying. Deutsche Bank would see an immediate—and significant—financial benefit.

Naturally, Deutsche Bank took Google Cloud up on this proposal. Google literally owns 25 percent of the internet! Their infrastructure is so vast and powerful and enormous. They were the perfect partner for an operation of this scale.

As a result, Deutsche Bank was able to move away from its legacy systems—and reduce the cost for maintaining them—as well as release the dragons from the basement.

Google and Deutsche Bank, partnering with Oracle, too, ultimately migrated 1,300 databases to the cloud. In addition, they upgraded and optimized thousands more so they could run their operation entirely on the cloud. The result for Deutsche Bank was the ability to eliminate all their former data centers (and therefore all the associated hardware maintenance costs). They increased functionality and reliability of their systems, at the same time generating cost savings amounting to hundreds of millions of euros per year.

SIX

When IT Works

Successful Digital Adaptations

A S WE SAW IN THE LAST CHAPTER, BEING READY
to adapt is crucial to ongoing success. But many well-in-
tentioned digital transformation and IT initiatives that
may *look* ready to go ultimately fail to produce the anticipated results. Some run over budget and end up too costly
relative to their deliverables. Others do not succeed at delivering their deliverables in the first place. What goes wrong in
these efforts?

It's a good question, and one we ask because none of us wants
that to happen in our organization. But rather than asking why

some things failed, let's look at why others succeeded. What are these other, more successful, companies are doing differently? For IT and digital initiatives in particular, what were the key factors that drove them toward success? What common factors do these success stories share? And, most importantly—what can we learn from these examples?

WHERE IT AND DIGITAL TRANSFORMATION MEET

There is a lot to learn in the three case histories we'll explore in this chapter—two of which I had the good fortune to be a part of—where IT and digital transformation worked.

Re-Envisioning Content: National Geographic

I feel fortunate that I had a front-row seat to the process of transforming both how content is created and how it is shared when I joined National Geographic to serve as executive vice president and chief technology officer in 2017. My dual roles were:

- to lead our product and technology team in delivering world-class digital consumer experiences
- and to design and implement an intelligent platform and architecture strategy, in tandem with our partners 21st Century Fox and Fox Networks Group.

National Geographic is a storied institution with a long history. Its yellow rectangle is a symbol of iconic photography, recognizable worldwide. However, as the twenty-tens unfolded, the organization was struggling. People were no longer buying paper magazines and there was a move away from cable television, so subscription revenues were down. Something needed to change.

My job was to bring this organization fully into the digital age through developing innovative products and technology, working closely with my boss, Rachel. I was charged with leading cloud migrations: Our team was responsible for making the company's breathtaking visual content more available to a large global audience—securely and effectively.

To accomplish this, we wanted to create new digital consumer experiences and technology that were in keeping with competitors' content, which was increasingly available online. But we knew that success required achieving a delicate balance, because we also needed to protect the great legacy of knowledge accumulated by this prestigious organization since its founding in 1888.

The previous CIO had attempted to implement this type of transformation. He had come up with lots of technical specs and architecture diagrams, which he presented to the team and which explained all the fantastic digital capabilities that could be available to them. But they didn't grasp his vision. And they didn't actually care, because they were all working away busily producing National Geographic content.

So, my first challenge as a tech leader would be to get my whole team on board with a process that required a lot of them: They'd have to stop doing what they had been doing and start doing something they hadn't done before. To overcome their understandable reluctance to change gears so completely, I would need to articulate a more engaging vision—a vision that was not about the technology per se, but about the impact we were going to have on our organization by implementing that technology: faster deployment of articles and photographs, more effective monetization, and greater reach enabling us to connect with new global audiences.

That sense of purpose would serve team members well when everything around them began to change. They were used to working in two distinct types of teams, physically separated from one another. In one part of the building, the traditional IT team focused on computing issues, while in an entirely different part of the building, the products team was trying to build things that their customers wanted. My first move was to unite them into a new products and technology team, a cross-functional team that had designers, software engineers, and the product people all working together. This meant changing roles and responsibilities, as well as many new hires.

The next step was changing how our IT worked—from systems to platforms. Up until that point, Nat Geo had been operating thirteen different computing systems in which editorial staff could create content. If the products people wanted to develop something new, they had to go into thirteen different systems to make their product operational!

Changing this would be our biggest technological challenge: We had to create a single cohesive cloud-based platform, so that all our photographic and written and video content was housed in one place. We met with several cloud providers. One, in particular, demonstrated a genuine interest in our business: Google.

Google's dedication to building a sustainable cloud paired well with our heritage in showcasing our planet's beauty. More importantly, Google was prepared to get hands-on and help us fix our app. They helped apply security patches, enhanced firewall capabilities, added identity and access control management, and offered advanced machine learning to enrich the archive's metadata. We were also able to take advantage of Google's extensive training and education for our team, which is essential for a successful deployment.

Finally, I initiated change in our operating model, so we could identify and use different metrics to define the success of the organization. As a result, we successfully re-envisioned how a major legacy company manages its content. The Nat Geo photo archive now supports over 400 million consumers a day on its digital channels.

Efficiency Through Innovation: Carlsberg

A decade ago, legacy brewing company Carlsberg found themselves in trouble. The Danish multinational, founded in 1847, was struggling to remain profitable in a world where production costs were rising, consumer tastes were ever-changing, and overall, there was a market shift away from beer and toward wine and spirits (or, indeed, to not consuming alcohol at all).

The company was spending more and more to sell to a diminishing market.

In the summer of 2015, Carlsberg's CEO, Cees 't Hart, announced an ambitious new seven-year plan. This legacy company would invest in becoming a truly digital enterprise, with the dual aims of cutting overall operational costs by a third and investing those savings into future growth. He dubbed this project Sail '22.

Before Sail '22, the company had been operating its global process workloads on legacy systems via an on-site data center. Now it would undergo one of the largest digital transitions ever to take place in the corporate world, migrating 100 percent of these workloads to the cloud.

This ambitious project was nothing to attempt in-house. The CEO was very aware of that—he understood that his company's expertise was in brewing, not in technology or cloud migrations. His team selected Accenture as their tech partner, to help guide them through the project.

In less than two months, Carlsberg and Accenture team members came up with a plan: If their cloud migration was to take place swiftly and with no disruption to business, they would need to contract the best and most qualified service providers. Microsoft Azure was their cloud platform of choice. Avanade, experts in Microsoft technology, would plan and execute the migration. And Zscaler would handle cybersecurity.

In the space of six months, Carlsberg migrated 1,300 servers and 650 critical business applications to the cloud. This meant they were able to shut down their own data center, while at the same time creating new efficiencies through innovative applications, such as placing internet-connected sensors on beer kegs to collect real-time consumption data for use in marketing campaigns. They also streamlined service delivery by using intelligent tools and increasing automation.

As a result of their Sail '22 digital transformation project, Carlsberg saw significant progress toward their KPIs, not to mention cost savings that resulted in cash return to shareholders through dividends and share buybacks. In March 2022, in an effort to continue building on that success, Carlsberg announced a new Sail '27 strategy, a company-wide initiative that builds on those efficiencies to help the company continue to expand its markets.

An AI Insurance Syndicate: Ki

In 2019, when I was working for Google Cloud, I was asked by my CEO to join a client meeting. That meeting would be with the CEO of Brit, one of the biggest commercial insurance brokers on the Lloyd's of London insurance marketplace. The CEO was not a client of Google, so at first, I was a bit confused about why I was being asked to attend this meeting.

Lloyd's of London was founded in 1688. It has dominated the insurance industry for over three hundred years. And although it has always had superior technical underwriting abilities and an edge in pricing and complex risk insurance, it didn't seem to me the type of organization that would even have digital transformation on its radar. I was intrigued: Of course, I agreed to attend!

The CEO was Matthew Wilson. He was just what I expected the CEO of a giant legacy insurance company to be—a charming, dapper, English gentleman. He got right down to business.

The insurance industry had failed for decades to transform in any meaningful way, he explained. They still used processes that were defined three hundred years ago and seemed reluctant to change. Paid consultants from the likes of McKinsey, IBM, and others had weighed in with insightful strategy suggestions, and yet, nothing happened. Wilson was frustrated.

Then he turned toward the fellow on his left and said, "This gentleman is my head of innovation. He's asked me whether I've talked to Google. He said I should ask you what Google would do. And that's why I've invited you here."

This was the start of us working together. To begin, I needed to understand how their business currently worked. To do that, I spent some time at the Lloyd's of London insurance marketplace, on the marine floor, where people insure assets such as ships and oil rigs. And, on my first day, I observed what, to me (in 2019), was the most bizarre scene!

This section took up most of the ground floor. The building was made of glass, open in the middle, with huge concrete columns supporting the structure and a giant central stairway of steel zigzagging its way up through the floors. In spite of the modern architecture, the marine section had the feel of an old boys' club, with carpets and lots of wood, crammed with tiny cubicles and people hunched over typing away at screens. Every now and then a head popped up.

A gentleman walked in with a thick folder jammed full of papers under his arm. He scanned the room, made eye contact with someone, and waved. Then he walked over to them and plopped his big pile of papers onto this person's tiny desk.

I was with Brit's CFO at the time, and I asked him what was happening. He explained that the gentleman who had just walked in had an asset that needed to be insured. He had come up to the marine floor to try to find someone with whom he could discuss his asset, so he could then get an insurance policy.

We walked over so I could observe. Interleaved within his pile of papers was a picture of an oil rig, as well as lots of charts and spreadsheets and other snippets of information.

The CFO explained that this is the typical process for insuring an oil rig. The man and the assessor will talk about all his data for a while. Perhaps they would go and talk to some other people too. Then, after more meetings and review and whatnot, in a few weeks or so, the gentleman will come out with a policy for insuring his oil rig.

I told the CFO that this seemed a very archaic way to compile and assess this man's information. He asked me, "Well, how would Google do it?"

I explained: We would want at least a thousand data points. We would look at every variable, from the strength of the steel used in the rig's construction, to the pressures in the drill-head, to the movement of the seas and the typical weather systems. And we would use all these variables to calculate our risk.

"We don't do any of that," responded the CFO. "Our system is all based on the knowledge and expertise of our assessors."

His assessors were, indeed, experienced and competent. A person who has worked for twenty or more years insuring oil rigs certainly has an implicit feeling for the effects of these variables. At the same time, it is difficult (if not impossible) for a single human to take into account the thousand or more variables that an AI can process—or to rigorously compare data from one rig with all its variables to data from someone else's rig with its unique variables. And it's even harder to ensure consistency between different assessors' evaluation of risk.

We could do better. So, we came to an arrangement for Google to work with Brit. I was the office of the CTO lead on the project. We would follow a data-led approach, eliminating document dependence and automating processes with algorithms.

Together, we built the world's first-ever AI-based insurance syndicate. Instead of human beings using their personal experience and assessments to estimate risk, our system input data. The result was a real-time and unbiased assessment of risk; potential clients no longer had to wait months for individual assessors to review the file and think about things.

Switching to a risk model using AI provided many benefits to the company. Not only did the AI system of plugging in data for an immediate response save time and result in standardization, but it also reduced scope for error. And, perhaps not surprisingly, having a system that required a smaller team with less paperwork to do led to reduced costs and therefore greater profitability.

They called this new AI-based syndicate "Ki." The company was the third-largest tech start-up in the United Kingdom in 2020. On its first day, it took in £550 million of investment. Last I checked, it was approaching £5 billion.

WHAT MAKES IT WORK?

Can we use these positive stories to deduce some of the key factors for ensuring success? I think we can. To get started, let's look at some of the elements that these three examples of successful digital transformations had in common.

Identifying the Problem

The first step toward solving a problem is acknowledging that you do indeed *have* a problem. If your business is foundering—you are losing clients, or struggling to be profitable, or always finding yourself two steps behind the competition—the first thing you need to do is to identify what your problem is.

In the case of National Geographic, the problem was how this legacy organization could remain relevant in the rapidly changing digital age. For Carlsberg, the problem was mainly financial: Production costs were too high to allow for profitability, so they needed to streamline and reduce costs. And for Brit/Ki, the problem was a bit of both: how to find new ways to operate in the digital world and to reduce costs by doing so.

Contracting Expertise

If one key to digital success is moving swiftly, then you need to have the very best people on your team. And I can tell you right now: The best tech people on the planet are not working for you.

They are working for the big tech companies, who provide them with a more stimulating and challenging environment (as well as pay them more).

So, in today's world, if you want to move forward with a cutting-edge and world-class digital transformation, you need to contract the expertise.

In the two projects I mentioned above that I was involved with, Nat Geo and Ki, we partnered with Google and Google Cloud. (Interestingly, these partnerships can work in both directions; for the first project, I was working for Nat Geo and we contracted Google; for the second, I was working for Google, and Brit contracted us.)

Similarly, our old Danish brewery, Carlsberg, assembled a team of specialists. They worked with Accenture as their consultants, Microsoft as their cloud provider, Avanade for tech support, and Zscaler for security.

Working with tech partners is absolutely key to implementing any meaningful digital transformation. Otherwise, you risk throwing endless funds toward your own in-house but outdated IT department. (We will talk much more about how to avoid "feeding the dragon" this way later in this book, as well as how you can gradually and sensitively repurpose your IT staff—and possibly your entire tech infrastructure too—to both save costs and move into the future.)

Vision

You may have noticed a common thread in all the successful digital transformations we have discussed in this chapter—such success requires getting the team on board. This is crucial, and it can't be accomplished by dazzling your staff with specs and speeds and how many servers you will be migrating to the cloud until their eyes glaze over. It requires the presence of an IT leader who can provide the team with a simple vision of what they are working toward and why. A leader who can give them something they can all buy into.

That was my challenge at National Geographic. I will explain in the next chapter how I approached that and presented my team with a vision that they both could understand and would buy into.

Carlsberg's CEO had a similar challenge, and he was very proactive with the vision he presented to his personnel. He explained what they were moving toward and how they would do it and what the timeline was—and he gave it a name, calling it Sail '22, something they could all rally around and work toward.

Likewise, working with Brit, we had a clear vision of our problem (and therefore our challenge), and what we would be moving toward: a highly automated system that we named Ki.

I write about these cases to demonstrate that the role of the enterprise IT leader has changed. No longer the engineer down in the basement, keeping the systems running and the network connected and writing software on demand, the IT leader of today (and tomorrow) needs to have a vision. They need to be able to communicate that digital vision to their entire team—not just to the people in the tech department who speak their language—and to rally their entire organization around it.

PART III

Transform

SEVEN

Befriend *the* Cloud

ARE THE INCREDIBLE DO-OR-DIE TRANSFOR-
mations we've been talking about really possible?

Yes, they are, because we are living in what is arguably the greatest meteoric revolution since the dawn of the digital age itself—the era of the cloud. This meteor is not just on the horizon—it's already here. You must adapt. Otherwise, if your enterprise is not already taking full advantage of all the capabilities of the cloud, you are already behind.

But how did we get here? And how can we harness the power of this moment?

To get a handle on that, let's first travel back in time to the year the cloud started to gain a foothold in the business world, around 2008, a lovely year you've probably tried hard to forget. Even though this was indeed a time of gut-wrenching financial loss, that annus horribilis failed to stifle innovation.

The "cloud" quickly became the newest (albeit controversial) buzzword. While some pundits paid it scant attention—considering it merely the newest trendy jargon referring to the internet—a 2008 article by *The Economist* sagely predicted that the cloud would "profoundly change the way people work, and companies operate."

The Economist proved to be very, very right.

That same year, Netflix became one of the first major organizations to migrate to the cloud. It shifted its resources from traditional physical data centers to the largest cloud provider, Amazon Web Services (AWS). One immediate benefit of this cloud migration was the cost savings associated with the closing of their physical data centers by strategic outsourcing: They were no longer condemned to feeding their dragons.

AWS's pay-as-you-go model of cloud computing provided several advantages. Previously, Netflix's own physical data centers needed to be capacitated to handle peak loads, in order to avoid the risk of any disruption to its streaming services—even though that level of usage was rarely attained. Now, Netflix could dynamically allocate resources, paying only for what it used. At the same time, it retained the ability to scale up its resources any time that demand increased, without the risk of outages.

Netflix's early migration to the cloud ensured its rapid success. Adopting cloud technologies and embracing platforms rather than systems allowed Netflix to take an agile development approach. It could deliver new features and updates rapidly, thereby securing its position at the leading edge relative to its competition. Basing itself in the cloud also provided additional security and resilience through many redundant systems located in various geographical locations, reducing (or eliminating) the risk of service outages caused by local failures affecting individual data centers.

MORE THAN THE INTERNET: UNDERSTANDING THE CLOUD

Netflix's story demonstrates the power of the cloud, but we still haven't defined what "the cloud" is. That's for a good reason—any definition of the cloud is, by definition, nebulous.

A report by the US National Institute of Standards in 2011 attempted a loose definition of the cloud: "a model for enabling ubiquitous, convenient, on-demand network access to a shared pool of configurable computing resources (e.g., networks, servers, storage, applications, and services) that can be rapidly provisioned and released with minimal management effort or service provider interaction." They clarified that with a list of the cloud's five "essential characteristics." A report by the European Commission the following year declined attempting any definition at all, instead outlining a list of eight "defining features."

I don't have a firm definition either, but would like to offer my own list, based upon the previous definitions and lists and also on my own experiences and values. Here are what I consider to be some of the essential defining features of the cloud:

- The hardware (e.g., individual computers, data centers) is owned by the cloud provider, and not by the users who interact with it via the internet.
- Data centers may be located in various parts of the world, providing backup, redundancy, and scalability of capabilities.
- Individual cloud providers pool their resources in order to be able to serve multiple clients/users, dynamically optimizing capacity by moving users' workloads around or between various computers or data centers.
- It is irrelevant to the user where the computers are or which computers are being used at any particular moment.
- Users can access their data, content, and software whenever and wherever they need it, and from a range of devices (e.g., desktops, laptops, tablets, smartphones).
- Usage is on demand and scalable, meaning users can rapidly modify the amount of computing power or

hardware they are using. To the user, this appears as having infinite capacity.

• Users normally pay according to their usage.

The trouble with defining the cloud is that it's not one thing. The term refers to an architectural principle in which computing resources are "virtualized," a network of servers located in data centers around the world and accessible by the internet, along with the software and databases that run on those servers.

However we define it, the advent of the cloud was a huge deal, similar in some ways to the arrival of electricity provided to users from a central source. Farms and businesses once relied on on-site electricity generators. However, about a century ago, individual users began to shut down these generators in favor of power sourced from more efficient, industrial utility providers via the grid. Today, business owners are doing much the same: shutting down costly and less efficient in-house computing and storage solutions and trading them in for cloud-provided capabilities.

In most cases, we still pretend to have our own dedicated computers, though these are really "virtual computers"—servers on the cloud partitioned by a software device called a "hypervisor." Each virtual computer behaves to its users as if it is an actual individual stand-alone computer, but instead of an application running on a specific piece of hardware, like it would in-house, it runs on a "logical" server, which may reside on a physical piece of hardware alongside other virtual computers, rather than on a physical server.

On the internet, servers respond to requests from users or clients. For example, a user queries a website, and the internet provides a response. In cloud computing, however, the servers are doing far more than simply responding to requests. They are actually undertaking functions and tasks for their clients, such as storing data and running programs.

So we see that the cloud does indeed *use* the internet, but it is much, much more than that.

The real power of the cloud comes from the many attributes of its shared servers around the globe. First is redundancy. Should one server be affected by a local outage (for example, a power failure), applications can continue to be run seamlessly from a different server located elsewhere. Redundancy also offers the safety of having multiple backups available in a range of locations should one server fail altogether.

A second attribute of the cloud is the scalability of its functions—and how that leads to cost savings. In essence, the cloud has democratized IT by giving users (like you) access to unlimited amounts of storage capacity and computing resources, paid for via a scale directly related to usage.

Let's say a business has three traditional servers running three different applications. The servers are designed to handle peak forecast loads, but normally two of the servers are being utilized at 40 percent each, and the third at only 10 percent. Even though these servers mostly run at well under half of their capacity, the company operating them still must pay the full costs of buying, managing, and maintaining all three servers. In large organizations with thousands of servers, this wastage can represent a high percentage of overall technology expenditures.

In contrast, with the cloud's virtualized servers, you can optimize all the different applications into a smaller number of physical servers. Each virtual server provides the resources that the application needs, with no wastage. Why would an online retailer want to pay for a server that they use mostly at 10 percent, but need to have available at 100 percent once a year for Black Friday? (And even so, risk outages due to it being overwhelmed?)

Finally, instead of investing in fixed data centers and condemning yourself to continue feeding your dragons for a few more decades, cloud computing allows agility: the ability to swiftly change direction or respond to new market dynamics. Rather than investing in hardware (and IT personnel), which may be hard to modify or change out or get rid of, moving operations to the cloud provides flexibility. If a software or system or platform is not working for you, you can quickly create another. Working on the cloud allows you to upscale (or downsize, or even halt operations) rapidly, without investing in (or warehousing) equipment.

An enterprise operating thousands of servers will welcome any improvement. Even a 10 percent improvement in efficiency may equate to millions of dollars of savings.

The first step is deciding what kind of cloud computing arrangement would work best for your company. You can choose among a number of different service models for cloud computing (in other words, what you pay for), and several types of cloud deployments (meaning where your data are stored). It's important to understand these as you plan your migration to the cloud. You want to be sure that you make the best decisions for your organization's own unique needs.

CHOOSING AMONG CLOUD SERVICE MODELS

So, what are you paying for when you contract services via the cloud? That depends on what you need, and *that* depends upon the balance between what, exactly, you plan to outsource and what you wish to create in-house. From there, you can choose a cloud service model that fits your organization's needs.

Your choice is among the four main service models in place today, which we'll cover in this section. Before we dive in, I would like to point out that though these categories are useful, there are not always hard dividing lines between them. The right service model for you might straddle the boundaries between these models.

Infrastructure as a Service (IaaS)

IaaS is the most basic of cloud services and can be thought of as essentially renting physical hardware (in other words, space on servers for computing and storage) from the cloud provider. Servers are partitioned via the hypervisor software, which separates the server into virtual computers.

Each client then accesses a virtual computer, which is one partitioned section of the server, to use that hardware to develop their own applications. The space they rent on the server behaves like a unique computer, which means the user has control over the operating systems and development frameworks.

The IaaS model provides on-demand service for the clients, allowing for dynamic scaling. Therefore, multiple clients will typically be using the same piece of hardware. You, as a client, will not necessarily know which hardware you are working on. The cloud provider may shift your workloads to a different machine in order to optimize functionality, without you even being aware of that. Examples of IaaS providers include Amazon Web Services (aws), Microsoft Azure, OpenStack, and Google Cloud.

Platform as a Service (PaaS)

PaaS is the next level of complexity in cloud services. As in IaaS, the PaaS cloud provider offers the client access to hardware, but the client also has access to the platform, which is like a set of software tools that can be used by the client's developers.

Having that access allows in-house developers to focus on building and managing applications rather than worrying about any of the infrastructure. In other words, your tech team spends their time building experiences and creating value, rather than feeding that old dragon and keeping systems alive. Some PaaS examples include AWS Lambda, Google App Engine, SAP Cloud, and IBM Cloud.

Software as a Service (SaaS)

Chances are that you already have personal experience with SaaS. This is software that you can access online, usually directly via your smartphone or web browser, and usually through a paid subscription fee. Examples of SaaS include Google Workspace, Dropbox, Trello, Slack, and Adobe Creative Cloud.

Unlike IaaS, which is used by network architects from your tech team, and PaaS, which is used by software developers, SaaS products are accessed and used directly by the end user. SaaS products require little or no specialized technical knowledge or training (or assistance from IT staff).

Function as a Service (FaaS, or "Serverless" Computing)

Of course, "serverless" applications still run on servers—but FaaS is sometimes referred to as serverless computing because those applications don't run on dedicated machines.

FaaS is a way for cloud providers to deliver back-end computer services to clients on an as-used basis. Rather than reserving and paying for a specified amount of bandwidth or a fixed number of servers, the client company is charged based upon their computing usage. It is in many ways analogous to your cell phone plan options: Rather than requiring a fixed fee for a monthly data limit, which you may or may not end up actually using, FaaS works more like phone plans that charge per minute.

FaaS is a more recently implemented service model. Some of its advantages are that it is usually lower cost, as you only pay for computing you actually use, and that the services are easily scalable on demand.

TYPES OF CLOUD DEPLOYMENTS

Once you've chosen a cloud service model, you're ready to decide what type of cloud *deployment* you need. This choice is all about where the cloud servers are, who has access to them, and who manages them.

Private Cloud

A private cloud is a server or data center that is dedicated entirely to one organization. The advantage of a private cloud environment is, of course, the security. However, private clouds are expensive to set up. In addition, it may be very difficult for an organization to replicate on their own the level and scale of services they would get via a public cloud provider.

A private cloud can be set up in a variety of ways. An on-premises private cloud is the most secure, but another option is infrastructure *owned* by a cloud provider, but not *shared* with other users. The more efficient private cloud solutions are those hosted or managed by a cloud provider (such as HPE, Dell, Oracle, or IBM).

Creating an on-premises private cloud may appear to provide significant cost savings and operational benefits to IT systems. However, it doesn't often enable change or innovation. The applications running on those virtual machines are no different from when they ran directly on legacy hardware.

Public Cloud

When people talk about "the cloud," they are typically referring to the public cloud.

The public cloud emerged after several companies applied the virtualization principle to infrastructure that they made available to third parties across the internet. Instead of a client company using their own servers, they could now "rent" space on off-site servers provided by these other companies, such as AWS or Google Cloud.

These servers are located in data centers in multiple locations across the world. Although different organizations share access to individual servers (partitioned into virtual computers), strict security measures are in place.

The public cloud provider maintains and runs the physical servers as well as the networking infrastructure that supports them. They are responsible for ensuring that their clients get the appropriate power, connectivity, and security.

The public cloud offers a compelling advantage. It makes the cloud provider's resources for running infrastructure, as well as their vast experience and expertise, available to business organizations of all sizes—few of which would be able to develop or support the same on their own.

Multi-Cloud

Multi-cloud deployment is where a user utilizes multiple clouds (public or private), selecting specific services from different public vendors according to their strengths. For example, they might use one cloud provider for computing tasks, another for data storage, and yet another for data analytics software.

The goal here is dual: Select specific cloud providers for the areas in which they excel, and avoid tying your whole organization to a single cloud provider.

Hybrid Cloud

In the multi-cloud scenario, described above, the various clouds are used independently. A hybrid cloud deployment is somewhat similar to a multi-cloud, but it means connecting between a private cloud and a public cloud, or two public clouds.

This model may work for organizations that run a private cloud but require additional functionality that their private cloud cannot provide. It can also work when the flexibility and scalability of moving resources between an on-premises cloud and the public cloud is a value.

Additionally, compliance requirements in some countries insist that personal data are held only within their borders and never sent overseas. This may render public cloud usage impractical and force organizations to also run a private cloud.

IS THE CLOUD *REALLY* MY FRIEND?

Yes, I promise you, the cloud is indeed your friend! Migrating to the cloud will enhance your capabilities while reducing expenditures. But working in the cloud does mean thinking of issues such as security and redundancy in a different light.

The younger generation, having grown up with cloud computing, may be very comfortable with it. However, some of the dinosaurs may be a bit suspicious about handing over their organization's operations and their data to "the cloud." Resistance to moving to the cloud is most often rooted in these concerns about security. This caution is warranted!

Other concerns I have heard relate to reliability and availability. It's true that, with more and more operations and data centralized, the implications of outages or attacks may rapidly turn global (and expensive!). Therefore, it is wise to be aware of the risks associated with cloud computing, and to understand how to mitigate them.

Cloud computing is powerful, but it does indeed come with a different set of risks compared to on-premises computing. I would not say that the risks are greater—in fact, overall, the risks are smaller and fewer—but they are different, and it's important to become acquainted with them.

Security Concerns

Few executives are comfortable with the idea of being 100 percent dependent on services that they cannot control. Why would anyone entrust their core systems and data to a third party?

First, it is important to understand that cloud computing overall is inherently *more* secure than on-premises computing. That is fundamentally because of how IP addresses work.

Any server connected to the internet must have an IP address. This number is like the computer's "telephone number" on the internet. If a hacker can work out what a server's IP address is, they can begin to attack it. A company's server located in a company's data center therefore makes for an easy target: The hackers know where the servers are and what they look like, because they are static.

In contrast, a server on a virtual machine in the cloud is dynamic. Its location moves around by the hour, or even by the minute, according to where the cloud services provider sees fit to maximize utilization and reduce the cost. Any bad actor planning an attack faces a challenge right from the start, because they can't even determine where that server is.

Still, there are valid security concerns about spreading company operations across a global digital ecosystem and opening up potential windows for a cyberattack. More operations in more places presents an expanded threat surface, which requires new or different means of safeguarding. And public cloud computing means sharing hardware: partitioning individual servers into virtual computers that are accessed by distinct clients, which leaves some users feeling uneasy.

While all these concerns are legitimate, cloud data are monitored constantly and protected by end-to-end encryption. With the redundancy implicit in storing data in the cloud, rather than on an individual device, attacks such as ransomware cease to be an issue.

The reality is that AWS, Google Cloud, and Microsoft Azure have access to more engineering and tech operations than your organization does. Additionally, they have deeper pockets and greater expertise when it comes to technology. By moving your operations to the cloud, you are taking advantage of the culture, processes, and platforms these giants have built over several decades, rather than competing with them for talent or resources.

As long as you use reputable cloud providers and consultants and continue to enact sensible security protocols within your organization, the level of security on the cloud is higher than almost any organization could attain on their own. Here are examples of some of the procedures that you can implement to help keep cloud operations safe:

- Require employees to log out of cloud accounts when not using them.
- Use multifactor authentication (MFA), e.g., asking for a password as well as a thumbprint or face scan. (Microsoft estimates that 99.9 percent of account compromise attacks can be prevented by implementing MFA.)
- Segment networks, so that all parts of the system are not accessible to all users.
- Use virtual private networks (VPNs) as standard practice.
- Conduct periodic cloud audits that focus not only on performance, but also that scan for potential unauthorized access, and verify processes such as access management, vendor management, and data backup and recovery.

To mitigate security concerns, many large companies do not switch entirely to a public cloud-only strategy. Instead, they create an architecture where certain systems are still housed on-premises and are connected to other systems in the public cloud. This is the "hybrid" or "multi-cloud" strategy I talked about above.

Systemic Failures

No system is perfect, and of course there may be rare and brief periods when the cloud *does* go down, and you don't have access to the services. This is not the same as losing your data—it is simply a temporary loss of access or functionality.

The worldwide Microsoft outage of July 2024 is an example of a very large, temporary, systemic failure. It was not caused by any sort of cyberattack. Rather, it was caused by human error: a faulty content upgrade by Microsoft's security provider, CrowdStrike.

CrowdStrike's Falcon security platform is intended to detect and stop breaches and attacks. However, in this case, the update was released with an error that caused Microsoft computers to crash.

The failures first started appearing in Australia on the morning of July 18, 2024, with Microsoft computers around the country inoperable, showing the "blue screen of death" and unable to be restarted. Australia's national cybersecurity coordinator reported "a large-scale technical outage" affecting payment systems at retail outlets as well as the National Australia Bank, and airlines were forced to cancel flights as their information machines at the airports went blank.

Outages quickly rolled across the globe, affecting everything from hospitals and emergency services to airports and trains, to news outlets and television stations, to banks and even the London Stock Exchange. International air travel was disrupted as thousands of flights were canceled around the world.

Although CrowdStrike immediately provided work-around steps, and soon thereafter deployed a fix to the defective software, it took days for computing systems to be restored worldwide and for business to return to normal. It's estimated that financial losses worldwide due to the outage total in the billions of dollars.

Obviously, there is no such thing as a computing system that is 100 percent safe from outages, failures, or attack. In spite of isolated examples like this, overall, the cloud is safer than your own computing and storage system. Here's an example:

When I was with T-Mobile, we partnered with Google Cloud. Our headquarters were in Seattle, so we used Google's western region for most of our activity and services. However, everything we were doing in the western region was being replicated in Google's eastern region as a real-time backup. In the unlikely event that Google's storage area network devices went down in the western region, we would immediately be able to continue operating in the eastern region.

It is not just Google that offers this type of redundancy. Any reputable cloud provider will have systems distributed over several geographical locations with multiple backups. Your data will always be stored in at least two places (usually more), typically broken up into blocks that are both replicated and spread across multiple servers. It is very, very hard to achieve that level of redundancy, and therefore risk resilience, if you don't use the cloud.

It is possible (and probable) that your cloud services may occasionally be inaccessible. Your Service Level Agreement (SLA) should guarantee a minimum level of uptime (such as 99.99 percent, or 99.999 percent). If you provider does not meet the guaranteed time, they may owe you a refund; if they repeatedly are not meeting their commitment, you may want to look for a different cloud services provider.

Planned downtime will be rare, but there will be terms about this in your cloud SLA that you need to be aware of. Planned outages usually will be programmed for low usage periods and will be communicated.

Temporary outages, while brief and rare, can be expected. Nothing is perfect. So yes, there may be fleeting moments when you cannot access your cloud services—but with the replication and backup of data, and the geographical distribution of redundant servers, your data will all be safe and there for you as soon as services are up again.

But, you may be thinking, what happens if our cloud provider disappears altogether?

This is very unlikely to happen—aside from the case where you organization violates the terms of service. Such a case occurred in January 2021, when Amazon Web Services canceled its hosting services for the right-wing social media website Parler for "posts that clearly encourage and incite violence." However, this is an isolated and extremely unusual example of a cloud hosting service suspending an account.

You clearly need to think through your transition to the cloud, relying on knowledgeable experts to advise on security management as well as the "what-if" planning for possible disaster management. The details of how to do this are far beyond the scope of this book, but here are some examples of general principles:

- Be meticulous about backups. While your cloud provider should be backing up all your data, you should also independently back up everything you have in the cloud in a separate location.
- Aim to use several different cloud services, so you don't risk becoming locked into one provider if you encounter problems.
- Avoid custom Application Programming Interfaces (APIs) developed by your cloud provider—although they may appear convenient, they will also lock you into their platform.

- Where possible, use open-source software that can be run on several cloud platforms, in case you find yourself in disaster recovery mode.

YES, THE CLOUD IS INDEED YOUR FRIEND!

You may still be wondering: With all the things that potentially can go wrong, is it really wise to hand most or all of your computing operations to the cloud?

Yes! You must adapt and evolve, because cloud computing is one of the most significant meteors to impact our digital world in decades. Those who harness its power will quickly leave the others behind. Many already have done this.

Overall, migrating to the cloud will:

- reduce your infrastructure, thereby lowering your overhead and overall costs.
- eliminate your need for "feed-the-dragon" style IT departments and free your tech team to work on projects that create value.
- give you access to infrastructure that might otherwise have been unaffordable.
- give you the ability to scale usage or services up or down rapidly, without having to build (or warehouse) infrastructure.

- allow you to operate from multiple centers or internationally, as your users can access files and applications from any location.
- help you to increase productivity, with multiple users working on documents or data simultaneously: no more sending different versions of the same files back and forth by email!

Yes, indeed there are security issues related to the cloud—as there are with any computing that accesses the internet. You simply need to think of security in a different way, remaining aware of the different risks you may be exposed to. Remember that, overall, the security provided by established and reputable cloud services providers is much better than what most companies are ever able to implement on their own.

Meanwhile, the opportunities are immense: With the cost efficiencies, the scalability, and the opportunity to easily change direction because you are no longer tied to your infrastructure, making the move to the cloud seems like a no-brainer to me.

MODELS FOR CLOUD MIGRATION

I hope that I have, by now, convinced you that it is crucial that you migrate most or all of your operations to the cloud! Refusing to do so might be the most dinosauric move of the decade.

Migrating to the cloud is a big operation. For a small business, planning the migration and executing it may take several months or half a year. For larger enterprises the whole thing could take a year or eighteen months, or even longer.

That said, keep in mind that during the early stages of the COVID pandemic, large organizations managed to migrate the majority of their operations to the cloud in a matter of weeks. Once the pandemic started, Etsy, having recently green-lighted a two-year Google Cloud migration, ended up completing the project in less than a month! And likewise for the Italian school system, which had not even been contemplating a move to the cloud.

Planning your cloud migration needs to be done properly. This is not a process where you can tolerate inefficiencies or errors. But once the decisions are taken and plans are made, there is no reason to move slowly.

HOW TO MOVE TO THE CLOUD

I offer here three main models for how to implement your cloud migration. They vary by whether you are simply moving your existing operations to the cloud without changing them or creating entirely new applications.

These models can be considered three end members. Of course, any one company's plans may contain elements of more than one of these models.

Lift and Shift

"Lift and shift" is also known as re-hosting: It means moving exact copies of your workloads from your on-premises servers to the cloud. You lift your applications, data stores, and operating systems straight from your own computing network, and then shift it all to the cloud.

It sounds like an easy fix, doesn't it? In my experience, however, lift and shift thinking can turn cloud migration into a disappointment, even a failure, because it doesn't allow you to take advantage of everything else the cloud has to offer. If all you are doing is moving your old legacy systems over, you won't benefit from the many cloud-native abilities that newer applications have, and therefore you'll miss out on opportunities to enhance and optimize your functions.

In fact, a 2018 report by McKinsey found that organizations using the lift and shift strategy didn't always realize the financial benefits of cloud migration that they had been targeting.[1] To me, that comes as no surprise: Simply changing where applications and data are housed doesn't allow organizations to fully benefit from the cloud.

---

1   Andrea del Miglio and Will Forrest, "Creating Value with the Cloud," *Digital/McKinsey: Insights*, December 2018, https://www.mckinsey.com/~/media/McKinsey/Business%20Functions/McKinsey%20Digital/Our%20Insights/Creating%20value%20with%20the%20cloud%20compendium/Creating-value-with-the-cloud.ashx.

As you plan your migration, you need to force yourself to think differently about your applications and systems. You need to think cloud-first. A successful migration to the cloud requires a radical overhaul of current strategy, processes, and operations. Which of your current operations are no longer delivering value? What new functions would streamline your services or add value to your products?

Improve and Move

A move to the cloud significantly impacts business and technical strategies, skills, and finances. Management needs to anticipate that it will require a much greater change both in processes and in approach compared to other technology-driven initiatives. It's important to assess carefully what, precisely, you are trying to accomplish by moving to the cloud—from the big picture to the details. That makes it easier to assess how, instead of just lifting and shifting, you can improve all your processes as you go.

Improve and move means taking your existing workloads and refactoring them to take advantage of the enhanced capabilities of the cloud. This model is an ideal solution when either the cloud environment doesn't support your existing workloads (so lifting and shifting is not even an option) or when your existing workloads need updating anyway.

For example, when I first arrived at National Geographic, the National Geographic Image Collection, a searchable database of eight million photographs, was running on an old server on-site. In the old system, tagging an uploaded photo required someone to physically view the photo, then write down all the terms or key words that might relate to its content. However, sometimes people would make errors in creating those tags—for example, someone might mistakenly call a cheetah a leopard. And then it could be years before someone discovered that the photo had been tagged incorrectly. Anyone searching for an image of a cheetah in the meantime would never see this particular photo.

Our goal was to move this entire searchable image collection to the cloud. We retained its core functionality, which was the "move" part of the migration. But for the "improve" part, we took advantage of Google's many cloud-native capabilities to enhance that functionality.

For example, we started to use Google's automatic machine learning (ML) in image recognition for tagging. We could run an entire catalogue through the pre-built ML tools. The ML would do all the tagging for us, with a very high level of confidence—and much faster than any human could do the same.

And yes!—the ML could tell the difference between a cheetah and a leopard, by analyzing metrics such as angles and aspect ratios of the face and the ears.

As a result, we had an application that did essentially the same thing that it previously did. However, we substantially improved it when we moved it, taking advantage of cloud-native capabilities to make it both more efficient and more accurate than it was before.

Remove and Replace

Remove and replace is an ideal solution when your current applications are not meeting your goals. In this model, you decommission those existing applications—don't even try to improve them—and purchase or create completely new applications that fulfill those same needs but that are optimized to take advantage of the cloud.

This model for cloud migration is also known as rebuilding. It may take longer and cost more than the re-hosting or refactoring models. However, in certain cases, the enhanced functionality and scalability of the new applications will merit the extra investment in time and money.

CHOOSING A CLOUD PROVIDER

Currently, approximately two-thirds of the world's cloud infrastructure is controlled by three companies: Amazon Web Services (AWS), Microsoft Azure, and Google Cloud. There are a number of smaller or niche cloud services providers as well, but success in the cloud environment is heavily driven by scale. And therein lies the benefit of working with the Big Three—they own so much of this infrastructure. Their marginal costs of adding resources to support you are relatively minor, and the smaller players simply cannot offer that.

An even greater benefit of working with the Big Three is that they have the deepest talent pools. This means that they can operate more efficiently, and they have a greater edge on innovation and in rapidly evolving technologies such as AI. They also generally have better security.

There may still be occasions when you want to partner with a smaller cloud services provider—say, if the app you run is very specialized and there is only one company that provides it. Or you might choose a smaller niche provider if there is one that simply stands out as an expert in that software—for example, companies like IBM and Oracle, which provide proprietary database technology as well as cloud services. From these examples comes a gradation of some that call themselves cloud service providers, but that are essentially very limited SaaS providers, such as Adobe, and more recently SAP, which now offers their ERP software running in the cloud.

When you select your cloud services provider, you are (hopefully) entering a long-term relationship—a partnership. It's therefore prudent that you first analyze carefully what your future goals are and what your long-term needs will be. Then you can review potential cloud vendors to see who can provide you the best value: services specific to what you need at the best pricing.

E-commerce company Etsy did this when they made the decision in 2017 to migrate their operations to the cloud; they undertook a rigorous method for selecting a cloud provider.[2] They employed a decision matrix that contained over two hundred factors, with which they could score the cloud service contenders.

Factors Etsy ranked possible cloud providers on ranged from those related very much to the bottom line ("best negotiated rates" and "risk of budget explosion") to those related to capabilities and support ("auto scaling support" and "locations of data centers"), and even to value judgments ("good to their employees" and using "renewable energy sources"). Each of these factors was weighted such that more important factors counted much more toward the cloud vendors' ultimate scores than minor factors.

Using this matrix, developed specifically according to their own specific needs and values, Etsy arrived at a score for each of the cloud services vendors they reviewed. Of them, Google Cloud Platform came out clearly on top, with a score 10 percent higher than any of the other prospects. This does not mean that

---

2 Mike Fisher, "Selecting a Cloud Provider," *Code as Craft* (blog), Etsy, January 4, 2018, https://www.etsy.com/codeascraft/selecting-a-cloud-provider; "Etsy," Google Cloud Customer Stories, https://cloud.google.com/customers/etsy.

GCP is necessarily the best cloud provider out there; it simply means it was the best provider for Etsy's specific needs.

Consider Working with Consultants

By now it should be clear that moving operations to the cloud is a significant investment, and that planning your cloud migration and choosing the most appropriate cloud provider (or providers) for your specific needs is no small task. This planning is crucial to getting things done right, on time, and on budget.

However, chances are that you do not have the cloud computing expertise to plan and oversee this major project in-house! I strongly urge you to seek outside expertise from specialists—people who have done this many times before—to assist you with this stage. While your cloud provider will provide advisors, it is wise to seek outside experience and guidance as well.

There are numerous specialist consultancies with expertise in this area. We've seen some of them in action in previous chapters, for example, when Carlsberg quickly and efficiently closed down their own data center and migrated 1,300 servers to the cloud, thanks to planning with the help of Accenture and Avanade. In addition to Accenture and Avanade, other multinational professional services consultancies include Boston Consulting Group, Kearney, and McKinsey. A quick Google search will find you many smaller, niche companies that specialize in cloud migration.

MOVING THROUGH THE FOUR PHASES OF CLOUD MIGRATION

Every organization will have a unique plan, and therefore its own unique process, for migrating workloads to the cloud. How this is done—and how long it takes—depends on numerous factors, such as the size of the organization as a whole and what types of functions the organization undertakes. For example, a major telecom with complex operations and multiple data centers currently running hundreds of incompatible systems will have a much different plan from a smaller local or regional organization such as a real estate agency with, say, ten offices.

Therefore, rather than providing you with specific "steps" for your cloud migration, I am dividing the whole process into four phases. The second phase—making an appropriate plan—is the most important part of this whole process. Please consider my recommendation to use specialist tech industry consultants for this stage. No matter how much you may love the tech people you have in-house, you want your cloud migration to be planned by people who have done this many times before and who have seen it all.

Phase 1: Audit and Assess

This is the "discovery phase." It requires working out what you have, identifying what the challenges or problems are, and determining where you want to be. Depending upon your starting point, this may include:

- taking an inventory of the existing systems and applications that you want to migrate to the cloud, paying attention to items such as costs, security status, and business value
- determining whether there are any issues, such as licensing frameworks or compliance requirements (e.g., how sensitive data are handled in various countries) that will need to be addressed
- working out any workload dependencies
- determining the KPIs, or markers of success, that you are working toward: What do you want to achieve, and how will you know whether you have achieved it?

This assessment should give you an idea of the scale and the scope of moving your workloads to the cloud, while drawing attention to where any special challenges may be.

Phase 2: Plan Your Cloud Migration

You now know what you are dealing with: what you have, and what you want. This is where you figure out exactly how to do it.

Specifically, will you lift and shift, move and improve, or remove and replace? This is the most important phase of all—you need to get this right!

Now is the time to determine what works in your specific case. You have inventoried all your existing applications in the previous phase. Now let's examine each, asking:

- Which applications are the most valuable?
- Which of these might be enhanced in the cloud without major redevelopment (i.e., lift and shift)?
- Which of these could be substantially improved by taking advantage of cloud-native capabilities (move and improve)? Examples of this would be moving to serverless data processing tools like BigQuery instead of traditional databases, or using ML like we did at National Geographic to automatically tag images with key words.
- Are there any applications that we want to retire completely and replace with new, entirely cloud-native software (i.e., remove and replace)?

Phase 3: Undertake the Cloud Migration

Now you are taking action, actually moving your workloads to the cloud. Exactly how you handle this will depend upon the project, the needs of the organization, the types of workloads it operates, and especially the scale of operations.

For smaller organizations with simpler operations—let's say a small real estate company or a law firm—a cloud migration will typically start by moving the databases. After making sure that those databases have migrated successfully, then the applications that access those databases can be moved over.

However, large enterprises with many complex systems will not be able to undertake such a simple migration strategy; they will need to separate the project into groups (Google Cloud calls these groups "waves"). These waves are defined by applications sharing common characteristics or interdependencies. In other words, it makes sense to move these applications together as a group, because they relate to one another.

Typically, in complex migrations such as this, the easiest or most basic groups are moved first. These are usually simple stand-alone applications or databases that have few or no external dependencies. The more complex groups—those composed of applications with many interdependencies—require more planning and can be moved later, perhaps after subdividing them into smaller groups according to function, e.g., inventory management or fraud detection, for the actual migration process. (In Google Cloud terminology, these "move groups" migrate in individual "sprints.")

Phase 4: Monitoring and Optimization

Surprise! Phase 4 does not occur after Phase 3. Rather, it begins as soon as you begin migrating your workloads. From the start, you should also be monitoring, making sure everything is running as it should (and that no issues or problems have appeared), and optimizing the operations. Optimizing includes:

- setting up systems for automatic monitoring of workloads
- improving scaling systems to improve efficiency and reduce costs

- increasing the redundancy of workloads through servers located in different geographical zones
- creating and enhancing disaster recovery plans

Monitoring and optimization continues once the migration is completed. You want to ensure that there is a process in place for ongoing reviews of the performance of migrated apps, so that the plan can be updated as the organization's level of expertise with the cloud grows.

THE NEW CLOUD MINDSET

The cloud computing era is evolving fast. Extracting the greatest value out of your cloud investments requires a different mindset. If you're a technology leader, that starts with you.

I'm not saying that it will be easy to shift your mindset so completely. It was a big leap for me to make when I first came to embrace the cloud over fifteen years ago. Back then, as the CIO for the UK charity Comic Relief, much of my decision-making was informed by the traditional systems strategy training I had received from IBM. But my first use of external cloud services showed me that there was a different way of doing things, one focused exclusively on functionality and tailored to my team's need to deliver maximum value to the business. That first experience with the cloud allowed me to play an even more important role in the organization's leadership as the driver of the business's digital transformation. And it led me to rethink my entire role as a technology leader.

A move to the cloud necessitates a change in attitude in three areas: access, control, and risk. I had to adjust my attitude and approaches in all of these as I became more familiar and comfortable with the cloud. I suspect you will find the same.

Access

When you first move to the cloud, one of the things that stands out is that *everybody* can access it. Back in the dinosaur age, when we kept everything in data centers, it was easy for the CIO to control who had access, because they were in charge of an actual physical thing.

This is how it was when I first started with Comic Relief: Everything was in the data center, and I controlled the data center. When somebody wished to install a new server, or launch a new product, I had to physically let them into the data center; otherwise, they could not get it done. The only people who had the keys to the data center were people who reported to me. I had absolute control!

Control

But once I embraced the power of the cloud, suddenly people on other teams could access our AWS cloud service, logging in via their Comic Relief accounts.

I was no longer in control, which meant I had to adopt a different approach.

Rather than restricting access, I now had to make sure that our people were on board. I needed to make certain that they understood how I believed the cloud should be used, and that they bought into my guidelines and values.

The best way to ensure this is to make a clearly defined architecture part of your flexible technology. Everyone on your team should understand how everything works to their benefit. That way, people don't need to deviate from what the company is doing, and, more importantly, they don't *want* to.

This clarity creates a new role for you as the IT leader. The more you move away from absolute control, where you physically restrict what people can do, the better you need to communicate with your team. You're no longer the enforcer, but the educator and informer, delivering a framework your team can buy into. In this environment, they will naturally align with the view that you have.

Risk

Back in that old world, with our on-premises data centers, a CIO could feel pretty comfortable about security risk, because it seemed like they had absolute control; after all, it was their job, and theirs alone, to control access. But were they really so safe with the status quo, or did they only feel safe because they were doing what they were used to doing?

I suspect it's the latter, and I speak from experience: One of the reasons that we moved to the cloud at Comic Relief was that there had been a huge failure in one of our two storage area networks within our data center. All our applications that had data on that storage area network were abruptly unable to access any of their files.

This was our one and only data center, and our main machine was down. We lost a week of productivity waiting for a new replacement disk.

Had those files been stored in the cloud, with multiple backups being created in real time on servers located in dispersed geographical locations around the world, the failure of one disk or one server would have been imperceptible to the organization. Operations would have shifted seamlessly to a different server, and there would have been no outage, no loss of productivity.

The lesson here is that there is a difference between perception of risk and actual risk. As CIO, it might feel good psychologically when you, personally, control all the locks and keys, and you know no one external can get in. But in this case, we had no scope for action when something went wrong *internally*.

Adjusting to the world of the cloud means acquainting yourself with what and where the *actual* risks lie.

EXPECT AND EMBRACE CHANGE

When migrating to the cloud, even the most experienced technology executives will have to learn new ways of doing things. You will too. You'll face new obstacles, like moving from a world where you controlled everything (and felt good about that!) to a world where, now, your tech partner controls everything.

But you need to remind yourself that this is a good thing, because they will do a much better job of that than you can. They have more money, higher skills, greater talent, and deeper technology. And, unlike you, managing all that infrastructure is their core business.

When the entire landscape shifts significantly, it seldom pays to repeat what worked in the past. The old rules just don't apply. And when it shifts at lightning speed, you need access to experts on the cutting edge, experts that can keep up with the pace of change. That's very much the case when it comes to the cloud—and even more so to the newest kid on the block, AI or Artificial Intelligence.

EIGHT

Incorporate AI
in Business

F THE DEVELOPMENT OF THE CLOUD WAS A GAME-changing meteor, the advent of artificial intelligence (AI)—technology that uses logic and mathematics to simulate human decision-making and reasoning—is possibly even more so.

Artificial intelligence has captured the imagination recently. However, the reality is that it has been with us in one form or another for a long time. You may not think of Google Maps and Apple Maps as AI, for example, but they are, and they've long been integrated into the daily lives of millions of people around the world.

When you use Google Maps on your phone, you are giving permission for Google to take information about your movements and put that in its system. It does this not for just your phone, but for many phones, from many users, and over time. Then it can examine for your location: Where are you heading? How many other phones are running Google Maps nearby? How fast are those phones going?

From those real-time data, it can calculate how many people are on the road and how traffic is moving. Then, using machine learning (ML), it analyzes past data—for example, who else has done that journey, what route they took, how long they took. The result you see is a succinct Google Maps recommendation: Here is your best route and here's how long your voyage will take. But AI is doing the thinking.

Similarly, many of us have used autocomplete for writing texts or emails. When we type "Just got back..." it might offer "from, home, to" as suggestions. Most likely, the suggestion will be exactly the word we're looking for. On our end, it's just a word. But the MLs behind apps like Gmail or Microsoft Copilot have scanned millions of documents to analyze texts so they can calculate, statistically speaking, when three words are typed in a sentence, what the next word probably will be. Some of these MLs are general, based generically on what they are seeing across emails or texts. Others are very individualized, based upon what you personally have typed (or even done) in the past.

Another example of AI in daily action is personalized playlists, such as suggestions from Netflix or Prime for what you may also want to watch. They come up with these suggestions by analyzing two dimensions: what everyone else is watching, and your personal viewing history. Perhaps they also home in on particular topics, genres, writers, or actors you've appreciated in the past. Or maybe they evaluate what you last searched on Google or even what types of retail establishments you have recently visited.

UNDERSTANDING AI: TERMINOLOGY AND CONCEPTS

As familiar as we are with using AI in our daily lives, it can be hard to talk about, because we don't all use, or even know, the same vocabulary. So let's make sure that we are all on the same page with terminology, and develop our understanding of what AI is (and is not), along with some of its subsets.

AI: Artificial Intelligence

AI is a broad term that covers all forms of computer-based reasoning, including neural networks, computer vision, and machine learning.

Rather than simply programming computers to do specific and anticipated tasks, AI teaches computers or computer systems to perform tasks, even if they are new or unanticipated. You can think of this like the difference between a robot and a human.

A robot, like a computer program, can only do what it has been programmed to do; it is not capable of problem-solving or creativity. An AI, on the other hand, simulates human intelligence.

Just like any child who goes to school to learn basic skills, and then perhaps on to university to learn specialty skills, the AI needs to be trained. The training familiarizes it with myriad data and situations. Some AIs specialize in specific types of data, such as text, code, audio, image, or video. Increasingly, AIs are becoming multimodal, and able to understand and operate across multiple data types.

The AI learns patterns and how previous solutions have emerged. They can accumulate more experience than any human brain could every process. Then, just like a person confronted with a new problem to solve, the AI uses the training and the experience it has gained—even to solve new problems, or to produce new creative works.

ML: *Machine Learning*

ML is one subset of AI. It involves building models and algorithms that look for patterns in large bodies of data such as images or text.

For instance, ML can be used to recognize features or objects in images for facial recognition, or for video security monitoring—it can tell whether the movement caught by a security camera was a human or a cat. We used ML in my National Geographic digital initiative to catalog and tag images.

Machine learning requires human intervention. The machine can learn—but it learns in the same way as a human student who has a teacher or private tutor helping them along, guiding them or correcting them when they make mistakes. For example, in image tagging, it may require an engineer to start the process by manually identifying features and classifying the data, and then to monitor the AI results and adjust the algorithms as needed.

Deep Learning

Deep learning is a subset of ML. Like ML, it can handle various data types such as images or text. However, in contrast to ML, it requires little help or input from humans. It therefore requires a lot more computing power to undertake.

Deep learning is like someone learning how to do something from scratch—without a teacher or tutor or even mentor. Deep learning uses neural networks, so it works much like the human brain works. It may undergo hundreds or thousands of training rounds, analyzing patterns and correlations from scratch—like the most dedicated of self-taught human learners. A deep learning AI learns from its own errors. It does not need human intervention to adjust and correct, the way classic ML does.

Generative AI

Generative AI is one of the newest applications of AI. As its name suggests, generative AI is artificial intelligence that learns from large volumes of data and information to generate entirely new content.

ChatGPT is perhaps the most well-known generative AI. Not only will it answer questions for you in full sentences, as if a human were responding to you, it can write essays or letters for you with only a brief outline of what you need. Other AIs such as DALL-E (yes, in reference to Salvador Dalí) can create "paintings" for you based upon any imaginary scene you describe in a short text message.

Generative AI is already being used in a wide range of business settings for applications such as:

- enhancing medical images (health)
- generating marketing text and images (advertising)
- accelerating the design process of new products (retail, manufacturing)
- writing software code (tech)
- enhancing, sharpening, and enlarging images (photography)
- recommending investments (finance)
- creating audio and visual content (the arts)
- and much, much more!

A Few Additional Acronyms

We all love acronyms, don't we? Here are a few more, which you can use to impress your colleagues at your next meeting:

IOT, INTERNET OF THINGS

This is the interconnection of devices, from the coffee maker that you would like to commence brewing twelve minutes after your morning shower, so your cup is ready just as you finish combing your hair, to the automation (both monitoring and response) of workflows on the shop floor.

RPA, ROBOTIC PROCESS AUTOMATION

Don't get flustered by the terminology here: No robots are harmed (or even involved) in the process. It's only the process that is robotic, meaning repetitive. You can basically omit the "RP" and just think of the "A," meaning Automation. RPA represents undertaking repetitive and rules-based tasks, such as moving files from one location to another.

IDP, INTELLIGENT DOCUMENT PROCESSING

IDP is yet another process that largely involves the automation capabilities of AI. It aims to reduce potential bottlenecks in efficiencies caused by documentation: for example, an order not being shipped because the supplier did not receive payment, or health care treatment delayed because the required records were not received. IDP automates the process of manual data entry—be that from paper-based documents or from forms or scans—so those data can integrate with other business processes.

CV, COMPUTER VISION

This is the technology that uses computational techniques to derive information from images or videos. Essentially, it scans images and looks for patterns, learning as it goes.

NLP, NATURAL LANGUAGE PROCESSING

This is the technology that machines use to analyze natural language, both as written and as spoken by humans.

APPLICATIONS OF AI

Many executives are drawn to AI because of the increased efficiencies and reduced costs they can achieve when machines take over tasks that you formerly paid human beings to do (the automation application). That's extremely valuable, but there are many additional (and possibly much more powerful) benefits that AI can bring to the business world.

Let's get creative and imagine the uses of AI for all sorts of things: things like enhancing brand development, improving customer service, enabling deeper engagement with clients, and allowing for greater innovation coupled with the development of entirely new products.

AI has broad applications and implications. These are growing by the day. By casting a wider net and becoming more aware of possibilities, you can unleash numerous potential benefits of AI.

Here are some of the ways that application of AI may benefit a business:

Automation

Automation is the classic example of what AI is for—taking over repetitive tasks from your staff. This not only reduces costs, but overall, it increases speed and improves accuracy. Examples of application of AI automation include:

- approving credit card applications based on rapid analyzation of multiple data points
- identifying cancer diagnoses from X-ray images
- accelerating insurance claims processing: for example, by processing claims submission forms or by analyzing and categorizing images of damages

The application of AI automation is not necessarily virtual (as in taking place inside a piece of hardware). Sometimes it's physical. It was in 2019 that I first witnessed the utilization of an AI-enabled robot by one of the larger retailers to stock its shelves—in a physical retail store! When the store closed for the night, they would let the robot out. It would scoot up and down the aisles, identify empty spaces on the shelves, and then go get the products to restock the shelves.

Optimization

Beyond simple automation, you can also use AI to analyze and improve your existing systems. An AI can analyze years of business data in a very short time. This can enable competent leadership to make decisions that are entirely informed by hard data, to a degree that has never been possible before.

For example, in the business or production departments, AI can analyze previous performance to determine what has and has not yielded results. It can detect deficiencies in current systems and identify issues such as bottlenecks in a manufacturing process, or areas where ROI is poor. And it can help detect problems very early on by detecting mistakes or discrepancies in systems, so you can deal with them before they become major problems or disasters.

Not only can AI analyze data much faster and more completely than any human, but it does so with much more consistency (without any human bias) and fewer errors. This deep and impartial insight into large amounts of data allows for maximum optimization of systems and processes.

Customer Service

You may have a love-hate relationship with the virtual agents and chatbots on a company's helpline! But there is no doubt that these bots drive business costs down, while accelerating customers' access to assistance: no more spending twenty minutes (or two hours) on hold, waiting to reach a human being.

When properly deployed, virtual agents also enhance the end-user experience. Their greatest value is with first-contact assistance, vastly reducing the time for customers to experience an interaction. For basic problems or requests, an AI virtual agent can direct clients to the appropriate information, and they are generally able to access a greater range of product details and other peripheral information much more quickly than most human agents.

For more complex requests, virtual agents save clients valuable time (and reduce their frustration) by categorizing their issues and directing them to an appropriate response, such as programming a personal email or a scheduled phone call with a real human being. More rapid servicing of requests or complaints results in greater customer satisfaction, as well as time savings and thus reduced costs for businesses.

Visualization: Making Customers Want You

While some products for sale are physical and definable, and everyone knows what they look like, such as a car or a shoe or a lakeside vacation home, others are individualized, and thus harder for consumers to visualize. That makes them more difficult to market, because there is no image for the potential consumer to latch on to.

This is where AI visualization tools come in. For example, in 2022 Ingka Group, the largest IKEA retailer, launched AI software that let potential customers scan their home, or a room in their home, to visualize what it would look like with IKEA furniture and accessories.

Similarly, if you want to sell your home, rather than go through the time and expense of cleaning it up and staging it with new furniture, there are AIs that can virtually stage it for you. They use images of your room layouts, virtually clean up the clutter, and add new furniture to create marketing images. Or if you are considering a new hairstyle, you can try it out first with AI.

Content Creation

While computers have long been useful for completing repetitive tasks, the production of new and original creative content has remained the domain of human beings. Until now. The arrival of generative AI has completely changed that.

For example: "Generative AI is a type of artificial intelligence that can create new content, such as text, images, or code. It learns patterns from vast amounts of data and uses that knowledge to generate new, original material. For example, a generative AI model could be trained on millions of images of dogs and then create its own unique dog images that never existed before. This technology has the potential to revolutionize various fields, including art, design, and even scientific research."

I did not write that paragraph—Google Gemini did! I did write the prompt, saying "write me a paragraph explaining generative AI," and Gemini did exactly what I asked. This illustrates how rapidly and accurately an AI can create new written content, from brief written explanations like this to essays and books.

As Gemini explains above, generative AI can also be used to create original images from written prompts, such as "image of a tall cedar tree with a dark smoky sky in the style of El Greco." It can also be used to create music.

Generative AI is in the early stages of being able to create full videos or films. It is still unable to generate more than very short original video clips. Currently, therefore, AI-generated videos are generally hybrid productions between humans and the machines.

In 2018, an AI called Benjamin wrote and produced the feature film *Zone Out* after compiling scenes from open-domain sci-fi movies. It collaborated with human actors to face-swap and voice-swap lines in the selected scenes. While an actual throughgoing narrative may have been lacking, and Benjamin's dialogue was mostly nonsensical, the film nevertheless featured at the London Film Festival 48-hour challenge.

In the 2022 short film *The Frost*, these roles were reversed. The script was written by a human, Josh Rubin, but an AI was used to generate images for the various scenes using OpenAI's DALL-E 2. These images were then animated into short clips with another AI called D-ID, adding movement such as lips moving or brows furrowing. Although films made by generative AI still leave a lot of room for improvement, this is a rapidly moving field.

One fruitful application of generative AI that is already yielding good results is in marketing. Applications like Google Search allow Google to understand what tactics make people click and buy—and it does a much better job than any human advertising professional! Couple that insight with Gemini's ability to produce original text and images, and you have an AI that creates more effective marketing materials than any marketing intern out there could ever do.

Monitoring

Network security is challenging, and the challenges only get greater. For example, if you are a bank with thousands of branches, all with PCs and servers and data centers, you know there are hackers out there constantly testing for any possible vulnerability. It is tough, if not impossible, to monitor absolutely everything.

A very smart cybersecurity company called Armis came up with a completely different approach: What if, rather than trying to watch everything to see if anything happens, you build an AI model that just quietly runs in the background and looks for anomalies? They could do this because they realized that every device—whether a Mac laptop or a server or basically anything connected to the internet—had a typical way of operating. It's as if each device had a fingerprint of how, and how often, it undertook various operations. So why not look for fingerprints that didn't match what was expected?

Armis trained their AI with a database of sixteen million internet-connected devices to characterize each device type's fingerprint. Then, a major bank contracted Armis to implement this technology. Its AI soon came up with a startling finding! All the TVs on the executive floor of the bank were behaving differently from the baseline of those same TVs everywhere else around the world.

They investigated, and found that all those TVs on the executive floor were connecting to an IP address in China! The TVs were recording all the audio and sending those conversations back to China. They had been hacked.

In-House Training

Computerized training involving videos or PowerPoint-style presentations, coupled with quizzes, has long been available. While these offer many cost efficiencies, they lack the interactivity of human-led training: Learning opportunities are glossed over or missed altogether because attendees cannot ask questions or request clarification or contribute their own knowledge or experiences.

AI can not only develop training programs, but it can also create interactive virtual trainers or assistants that can answer questions or provide other on-demand support. As for any AI, the effectiveness and accuracy of the training, and of the AI's responses, relate directly to the data used to train the system—but there are many new companies out there that specialize in helping businesses develop AI-based in-house training.

NEW PRODUCTS, NEW NICHES

Speaking of new companies and new products...the emergence of generative AI has allowed for the generation of all sorts of new products. Training is but one example.

The list of new niches that take advantage of AI's abilities is long and includes NLP, human-like engagement, and analysis to create training products for sale, such as language courses and tutoring assistants for students. And it goes beyond screen-facing applications. Since 2019, the Eastern Pioneer Driving School, on the outskirts of Beijing, China, has been using AI instructors in cars to teach students to drive, with only light monitoring by a few instructors in a central control center.

Photo editing is another area in which AI is being used to develop entirely new products for sale. In the past, the maximum size that a photo could be printed was limited by its number of pixels. Products such as Topaz use ML and generative AI to interpolate and generate new pixels, as well as to analyze images to discover reasons for blur (for example, is it due to subject movement, or camera movement, or poor focus) and to undo the blur and sharpen the images.

We are still in the earliest stages of using AI to create new products. The coming years will reveal much in what human creativity and vision coupled with the power of AI and its many tools may have in store for us.

Predict Market Trends

Maybe AI itself can tell us what's coming. After all, it is able to predict other things, like future customer behavior and needs, as well as changing market trends.

AI can do this by analyzing customer behavior in real time, which helps leaders direct and generate very targeted marketing campaigns and present them to the right demographic groups at exactly the right time. (If you have ever purchased anything on Amazon, you will have seen this type of AI in action, with the recommendations and upsells the software placed right in front of you.)

This helps with business planning, both for scaling the production of current products in anticipation of their future demand and for the generation of new products.

Collaboration with Humans

AI can also collaborate with humans in a number of ways. AI virtual assistants have been helping us for well over a decade now, both as dedicated general (and named) assistants like Siri and Alexa, and also in less obvious and quieter supportive roles.

One example is in decision support. If you have ever wondered why Apple's call center customer support is so good at providing personalized help, it is because their agents are constantly collaborating with AI. A call center operator talking to a customer on the phone can bring onto their screen what is called an "agent toolbox."

The agent toolbox is an AI that is listening to the conversation. In real time, it feeds any information that will be helpful to the agent onto their agent screen. This could be general information about the product the customer is calling about, but it also may be answering very specific questions about the customer themselves: What products do they own? When was the last time they called? What product are they using right now? Have there been any faults identified or recalls on that product? Is there any incompatible software installed on that customer's product?

AI allows Apple agents to do such a great and personalized job because this real-time stream of information allows the agent to very quickly understand the problem and find a solution.

Another example of AI collaboration is in the legal field. In the past, lawyers would have multiple associates reading contracts for them, providing summaries and looking for problem areas. Now they can have an AI do that and more. The AI can also feed all those documents and contracts into a single model, trained by looking at previous contracts as well as their outcomes. The AI can retain and analyze far more information than any human team, and then flag certain clauses or terms that may have led to unforeseen costs or to litigation in the past.

These examples illustrate just a few current possibilities—there are many more, and AI is advancing incredibly quickly. That's why it's important for all business leaders—not only those on the tech team!—to become familiar with this rapidly changing technology, both its power and its failings. Armed with that knowledge, you can start to creatively plan how to harness the power of AI to position your own organization at the leading edge of your field.

NINE

Create *a* Forward-Looking Action Plan

W HILE I CAN OFFER EXAMPLES OF SUCCESS-
ful transformations, I cannot make the perfect plan
for you. Your organization's digital transformation—
whether it's how you undertake your technical upgrades
and migration to the cloud, or your incorporation of AI into
your business—will be individual and unique. Your plan needs
to be tailored both to what your organization's capabilities are
today and to what your vision is for the future.

However, I can help you to *plan* your plan! Let's start by looking at the elements you need to consider in order to set yourself up for success. Then we'll talk about how to actually get started with an appropriate plan.

SIX DIGITAL SUCCESS FACTORS

There are countless factors that may contribute to the success—or failure—of your digital initiatives. However, over my years working with a range of different organizations, each with different abilities and requirements and goals, I have distilled my list down to six factors that I consider to be of supreme importance in planning any sort of digital transformation.

Flexible Technology

When designing a technology platform, you need to consider four foundational attributes: scalability, flexibility, performance, and cost. In an ideal world, every system or solution would support all four of these. In the real world, however, this is not usually possible. For example, a system that is scalable *and* is flexible *and* has high performance would also be costly.

It's simply impossible to achieve a perfect score in all four categories at once. But there is one you absolutely must not miss—*flexibility*. Without it, your business model, objectives, and direction cannot support rapid change, which—as we've discussed—is crucial in this day and age.

With the information provided in this book, you have in hand the keys to achieving flexibility. Key number one is shifting away from reliance on systems (meaning both your hardware and your software) that are difficult or impossible to update and hold back your ability to innovate, and switching up your operating model by turning to one that consists of platforms and the technical products they support (as detailed in Chapter 4).

By no longer tying your operations to fixed, systems-based technological architecture, you gain the flexibility of rapidly changing directions, or even completely halting an initiative that isn't producing results and initiating something else. Flexibility gives you the opportunity to try different things out and experiment.

Key number two is, of course, migrating your workloads to the cloud (see Chapter 7).

Experimentation

Now, how to begin?

First, let's look at what you shouldn't do. When I worked as vice president at Agency.com, a leading digital consultancy, I was responsible for establishing new client relationships in our London office. I was quickly struck by the number of clients who wanted to design and build digital "things" without ever testing their ideas or getting feedback from users to find out whether their concepts actually work.

You don't have to—and shouldn't—work with that kind of uncertainty.

Enter experimentation. Experimentation gives you a clear answer because it refutes, supports, or validates your hypothesis. And its use is not limited to scientific laboratories! Digital organizations can use experiments to garner insight into the cause and effect of particular factors or elements of their plan.

In order to achieve total optimization of a platform, testing with end users is critical. In digitally mature organizations, everything they do is tested, from whether or not a product feature is necessary to what color the "order" button should be.

This type of experimentation can be undertaken by what is called "multivariate testing" (MVT). Multivariate means that several variables are investigated at the same time. Typically, multiple options of several different features—such as the color of a button and placement of features on a web page—are all tested alongside each other, by rolling them out to a small percentage of users.

For websites and mobile apps, MVT works particularly well. Interfaces, content, and flows can all be subjected to rapid testing—provided that the program is supported by the right platform.

Measurement

Experimentation and measurement go hand in hand. Measurement is the numerical quantification of what you are testing. Measurement is important because it gives you the results of your experiments and guides your future actions.

Measurement is nothing new to those of us in the business world. There are numerous metrics that we use frequently enough that they have earned familiar acronyms: key performance indicators or KPIs, cost per acquisition or CPA, and net promoter scores or NPS.

During my career, I've become a little obsessed with KPIs. Knowing which targets to chase helps to focus teams and creates an effective framework of performance assessment for any initiative.

Using KPIs successfully hinges on two premises: First, the KPIs must be set appropriately, and second, the organizations must actively work toward them. Tight, measurable KPIs help people in multidisciplinary teams orient themselves and align properly.

KPIs also pinpoint problems early: A new product that fails to meet a particular KPI is an oracle of caution. High-performance companies shut down experiments and products that don't meet their KPIs, whereas low-performance companies simply "hope" that things will improve. In the digital world, there's no escape from data. KPIs are your friend.

Extreme Collaboration

Now, let's take a closer look at where the ideas you are experimenting with and measuring come from. They should come from everywhere—people throughout your organization know things you don't and can offer critical insights! But how often does a "lowly" worker have a great idea, or possess some unique expertise, but the opportunity is lost because he or she has no means to communicate that knowledge to the higher-ups? Due to the hierarchical structure of most businesses, such communication breaks are common, and countless opportunities like this go unheeded every day.

But what if teams could work in an environment free from hierarchy, title, or role? Where personal interactions were based upon needs and outcomes, rather than on each person's status?

What if requests for information or for tasks to be completed were responded to right away? So you could immediately continue on with your energy and momentum, rather than wait days or weeks to hear back from a colleague?

There is another way. It's called extreme collaboration, or XC—it's the idea that teams are formed based upon who has the best skills for the given mission. Their position or status in the organization does not matter, only their skills, what they can contribute to the project. Mobilized around clear KPIs and objectives, extreme collaboration thus transcends traditional organizational boundaries.

The XC concept considers not just the level of communication, but also its speed. It's easy to see why this is important; for example, any type of written correspondence—be that writing up requirements to the IT team to support a new initiative, or something as simple as drafting an email asking someone to send over a file—slows down the speed at which interactions occur, thus dulling the acuity of any team and slowing its momentum.

Back in the mid-1990s, NASA's Jet Propulsion Lab (JPL) became the pioneers of the extreme collaboration way of working when they formed a group called Team X, with the aim of improving the speed and quality of mission proposals. The idea was that creating an environment that is both electronic and social in nature would maximize communication and information flow.

They would achieve this by having team members work in a "war room" environment—working physically together, in the same room. Team members used a variety of computer technologies, as usual, but not at separate desks or in separate offices. Now, not only were team members staring intently at their screens, but they were also seeing and overhearing what other team members were doing. Just as hearing your name across a noisy room triggers you to turn your head, hearing a conversation with words related to his or her project, such as "power" or "propulsion," might prompt an engineer to leave their computer screen and pop over to join that conversation.

Did it work? Well, how long do you think it might take a NASA team to complete a space mission design proposal? By employing the concepts of extreme collaboration, Team X was typically able to complete a design proposal in three sessions of three hours a week—a total of nine hours!

Even though this full model of extreme collaboration—physically assembling your diverse team members in the same room—may not be attainable in many business models, aspects of extreme collaboration may still be implemented. Virtual workrooms that incorporate many aspects of social media, especially including real-time or near-real-time communication (as well as its addictive nature), can mimic many of the attributes of NASA's Team X war room.

My first brush with extreme collaboration occurred during my first weeks at Apple. An unexpected phone call from someone who called himself John was both a mini interview (covering everything from my technical understanding of video to my attitude toward animals) and an invitation to join a black (or secret) project that he was running.

My first call with this team was remarkable, a true testament to extreme collaboration at work. There might have been a hundred people on the call—but I don't know, because no one took the time to introduce themselves.

The sole focus of the meeting was on a set of numbered actions, each paired with a responsible person who was called upon to share their progress. It turns out my invitation to join the team was due to some trouble they were having meeting technical requirements. While intimidating at first, this laser-sharp focus on objectives and progress was refreshing. It was a manifestation of a "no-nonsense, no failure" attitude, taken to a degree unlike anything I had ever seen.

Familiarizing yourself with the concept of extreme collaboration and adopting whichever elements of it that you can within your organization stimulates creativity and speeds up timelines, both of which are fundamental factors for meeting with success.

Agility

Agility, in an organizational context, is the ability for an organization to quickly *recognize*, quickly *understand*, and quickly *react* to events such as market signals, market conditions, or changes in the environment. All three components of these must be in place, or the organization cannot succeed in being agile. Here is what happens if one of these components is missing:

- An organization that cannot quickly *recognize* that something is changing will not even realize (or at least not realize in time) that they need to act quickly and change something. Much of the ability to recognize comes down to monitoring.

- An organization that quickly recognizes changes but doesn't quickly *understand* what they mean will know that they need to act but won't know what to do. They cannot make a meaningful plan. They cannot react.
- An organization that can quickly understand changes but cannot quickly *react* to them will miss the best opportunities, and likely lose out to competition. This most often comes down to archaic organizational structures— those that are either too heavy on paperwork or require too many meetings before any decisions may be made.

When I was at Comic Relief in London, we were running a fundraising campaign that engaged millions of people. Every day, the marketing director and head of campaigns, Michele Settle, reviewed our KPIs from the day before, assessing the effectiveness of the promotions and providing marching orders for the day ahead.

As she surveyed the reports, looking for traffic, conversion, donation amounts, and press coverage, we would listen, poised and ready to act on any initiative she identified. Then, multi-disciplinary teams would mobilize quickly to change content, or update functionality, or kick off entirely new work streams based on operational performance.

Our flow, therefore, was that Michelle would quickly *recognize* issues, we would all work together to *understand* them, and then the appropriate team would *react* quickly and change whatever needed changing. This is an example of agility done right.

Absolute Focus on a Customer's Needs

This final factor is the one I think of as the magical factor. And it's the one that almost no tech team pays attention to—an absolute focus on a customer's needs. (The customers are, after all, the ones who pay all our bills!)

Having the most flexible technology, the most extensive experimentation, the clearest KPIs, the most extreme collaboration, and the most seamless agility all mean nothing if these actions are not directed toward solving problems and creating value for consumers! Focus on our customers' needs should be the North Star for all sections of a business organization—and that includes the IT team.

Steve Jobs famously said that he didn't believe in asking consumers what they wanted. Why not? Because it wasn't the customers' job to know.

And Henry Ford reportedly said, "If I'd asked customers what they wanted, they would have told me, 'A faster horse'!" (Henry Ford did not actually say this—but nevertheless, the quip illustrates my point. Ford's customers, or Jobs's customers, do not have the vision to understand the *possibilities* of what they might want. That job lies with the experts, the visionary tech leaders.)

Absolute focus on a customer's needs does not mean "the customer is always right." In fact, it is the opposite. People generally don't know what they want until you show it to them. That's why I rarely rely on market research. Our task as tech leaders is to read the things that are not yet on the page.

We incorporated this concept of focus on the customer's (future) needs in the digital transformation I led at National Geographic, when we redesigned the popular "Your Shot" photography community. In order to visualize what a redesign *might* incorporate, we first had to understand the needs of the community members.

Merely asking the audience what it wanted wouldn't have yielded the appropriate results; members of the community often talked about it simply as an online forum for photographers. Instead, we worked with several outside practitioners who observed and interviewed "Your Shot" community members as they tested many different features. From that experiment, we determined the users' needs to be:

- showcasing their work to enhance their reputations
- connecting with others to educate themselves and improve their skills
- being inspired by the work of others to help motivate them in their own work

The ultimate result was a new version of "Your Shot" that delighted the audience and allowed the team to quickly grow its community to over one million members. Every aspect of the redesign—from the vocabulary to the color palette used, to the introduction of new features like a discussion forum—was informed by the true needs of the audience.

COMING UP WITH YOUR PLAN

Before you actually start to do anything at all, you need to determine two things:

- Where are you at right now?
- Where do you want to be?

This first question addresses everything about your current operations, as we explored in the earlier chapters of this book. Which dinosaurs have you already uncovered? How far into the adaptation process are you? Are you already evolving?

The second question asks you to step into the visionary role of the new IT leader—what, exactly, is your vision for your organization in the future?

Now, assess the gap between your answers to the first question and the second. How close have your efforts to date taken you toward where you want to be and what you want to achieve? You may be nearly there or you may have a long way to go; either way, you now know what you need to do, and you've learned something about the scale of what you need to achieve. Understanding this puts you firmly in the first stages of creating an initial blueprint for coming up with a plan.

Planning is the most important part of any big project (including for writing a book!). If you get the planning right, then the doing actually becomes fairly easy.

However, as noted, I cannot make your plan for you. I can't even make a list of elements that your plan needs to cover, because every organization comes in at a different starting point and with different goals. That said, I don't advise going it alone; to explore these two questions sufficiently, you will need help from people who are experts in the field of digital transformation—people who have done this many times before.

GET THE BEST TECH EXPERTS ON BOARD

I have encountered many tech leaders who are very resistant to hiring consultants or working with partners. They usually have a simple line of explanation for me, often coming down to cost.

"It's just paying for other people's computers," they tell me.

"Yes," I reply, "but those other people are smarter than you."

It is not what the average IT leader wants to hear. But it is an overall truth. Working with consultants is not just paying for their computers. It's paying for their smarts and their experience. And the smartest people work for the tech industry: for Amazon or Google or Apple or IBM, or one of the many specialist tech consultancies. (Not only do those companies provide techies with a more stimulating work environment—they also pay them a lot more!)

Planning your digital transformation, migrating workloads to the cloud, examining how to harness the power of AI—these are not small moves! Your plan needs to be tailored specifically to your own organization's current capabilities and future needs. This is big and important (and risky). You need to find the very best people to help you make this plan. You need a systems integrator.

What Does a Systems Integrator Do?

The term "systems integrator" (or sometimes a system integrator, or control systems integrator, or just the integrator) usually refers to the person or people who are working with you on your tech planning, but it can also refer to the entire company you are partnering with. Accenture and Infosys are two well-known examples of systems integrators.

Systems integrators are specialists in the field of digital transformation: That is all they do. A systems integrator will have planned and overseen digital transformations for many organizations similar to yours in the past. They are up on current and future trends in the tech world and can visualize settings and solutions that your own tech team and leaders will not be aware are even possible. While most systems integrators are generalists, who can apply their knowledge and experience to any industry, some are specialists in fields with unique needs or practices, such as manufacturing, or the oil and gas industry. Each industry has its own unique challenges, so finding the right systems integrator to partner with is essential—especially for large organizations with many complex, interdependent workflows.

Choose carefully, because you'll be working closely with your systems integrator; they will guide each phase of your digital transformation plan, helping you to integrate your hardware and software and communication protocols in a way that leads to new efficiencies and takes your organization into the future. For most organizations, the two biggest components of the plan will be a migration of workloads to the cloud, and an assessment and plan for harnessing the power of AI into workflows.

Don't try to go it alone. I guarantee you, you won't have access to the same human expertise within your own organization that you will find in companies like Kearney or Accenture or Avanade or Google.

Clear Vision

With the right partner, a tech leader can focus on creating a compelling vision that is grounded in reality and executable. That vision must be built on top of a new architecture that leverages the power of the cloud and shows how the organization's IT will be transformed.

The systems integrator will get to know your organization inside and out, so they can help at every step of the way. They will be involved in the initial audit and assessment of your current operations. They will help you make your digital transformation plan, guiding you to determine things like who is the most appropriate cloud provider for your business, what is the most appropriate platform for your computing needs, etc.

A good systems integrator will customize all their solutions both to your industry and to your organization's needs, ensuring that new systems adhere to all compliance regulations and meet security protocols. They will assist with monitoring and optimization as your digital transformation proceeds. I cannot stress enough the value in having a single specialist person or team that becomes intimately familiar with every aspect of your organization and your needs and that guides the project from start to finish.

All enterprise-scale digital initiatives are undertaken with consultants overseeing—there is simply too much at risk to do otherwise—and you have lots of options. Many majors partner with Google—not only as cloud services provider, but for expertise in directing their digital strategy (such as National Geographic, Brit, and Deutsche Bank). And there are other systems integrators out there. Carlsberg contracted Accenture and

Avanade. Even the mighty multinational ERP software provider SAP used Accenture to plan its cloud migration!

STRATEGIC OUTSOURCING

One of the greatest hurdles that enterprises face when even *considering* a major digital transformation is dealing with the question: What will we do with our old legacy systems? We have so much invested in our data centers, our software, our IT team. What do we do with all that?

There is a solution, my friends! It's called strategic outsourcing. You are likely familiar with the concept of business process outsourcing, where you transfer some of your business processes to a third party in order to reduce costs or increase efficiencies. Strategic outsourcing, in contrast, is transferring your actual tech systems—from ownership of the infrastructure to its management, and even to your people, as we discussed in Chapter 5— to a third party.

Strategic outsourcing of legacy systems to companies like Accenture or Google or IBM or TCS has been a very effective mechanism for reducing costs and driving better tech outcomes. In the case I outlined in Chapter 5, Deutsche Bank outsourced the entirety of its tech operations to Google. This is a powerful example of how strategic outsourcing works. Let's go back and look at exactly how that happened.

Deutsche Bank had been struggling with high IT expenditures but few advancements. But they are a bank, not a tech company. A top global tech player, Google, knew they could run Deutsche Bank's tech operations much more efficiently—so they outright purchased Deutsche Bank's data centers, taking over their people as well, and contracting those tech services back to Deutsche Bank.

This was an immediate benefit to both companies. Deutsche Bank's tech expenditures were immediately halved. And on top of that, their tech needs were now being managed by the largest and most capable tech company in the world: They could get back to doing what they actually *knew* how to do well.

It's not only the few big tech companies like Google that will take over a data center. For example, UK retailer Marks and Spencer outsourced their IT to TCS, Tata Consultancy Services. TCS is a global consultancy, but it has strong roots in India, which means they can undertake a lot of functions much more cost-effectively compared to business operations in most other parts of the world.

Just as for the Google/Deutsche Bank example, TCS took over ownership of Marks and Spencer's UK data center, as well as their people: They all became TCS employees. TCS could then use its efficiencies of scale, as well as the fact that it has data centers operating around the globe, to determine which of its operations it would undertake there in the old Marks and Spencer data center and which it would undertake in its Indian data centers or in data centers housed elsewhere.

By doing so, TCS could now contract services back to Marks and Spencer at a profit. At the same time, the costs to Marks and Spencer—after ridding itself of its data center and all its associated costs (in other words, its dragon and the dragon-keepers)—were reduced.

By shedding their tech operations, global corporations can move their attention back to where their expertise lies and how they generate profit, be that banking or retail or brewing or transport. Leaving the tech functions to the tech experts not only allows businesses to focus on what they are good at—it also relieves them of the headache and hassle and cost of trying to keep tech functions at the leading edge in a rapidly changing world.

Look Up: Always More Meteors *on the* Horizon

Although I come from Europe, I have spent nearly all my professional life in America. It's been interesting to me to observe how local culture influences both business culture and business practices. The traditional hierarchical business structures common in Europe may have been very useful in their day. However, in these rapidly changing times, they are an inhibitor to success.

Rather than functioning in a world where we see "organizations as machines," we need to think more of "organizations as organisms." A machine operates based on presets and predictable factors. An organism, however, lives and breathes and changes based on environmental conditions, responding immediately to change.

Successful digital transformation is, by nature, far more attuned to this concept of organizations as organisms. If you try to think of your own organization in this way, it helps to set the stage for flexibility, agility, and market readiness in light of uncertainty and change.

EMBRACE CHANGE

The meteors will keep coming. Rapidly advancing technologies mean that the bombardment of changes will only accelerate. The challenge of business leaders this coming decade will be to keep ahead of change (or at very least, keep pace with it).

To be truly ready, you must accept some basic truths we've been exploring throughout this book:

- Those who fail to embrace digital transformation will soon become irrelevant.
- Systems will drag you down. Your future is in platforms and products.
- The cloud is your friend. Rapid innovations in AI are rapidly changing our world. Embrace them both.
- Running platforms based in the cloud rather than point-solution systems in data centers absolutely changes the roles of your tech personnel. These mandates change in the relationship between IT and

business, and especially in the role of the IT leader
or CIO.

What do all these points have in common?

They are all about change. Dinosaurs don't like change. They avoid it and ignore the signs of it. Meanwhile, their competitors are continuously preparing for it.

Transformation within companies involves drastic foundational change, which may entail restructuring services, operating models, and objectives. It is a total overhaul of "business as usual." The dinosaurs can only lose here. Agility, adaptation, and evolution are our only solutions.

The accelerating pace of change makes it increasingly challenging to be a tech leader today. The job is certainly not for everyone. However, for those who thrive on challenge, these changing times can also make our work both a lot of fun and rewarding!

ANTICIPATE THE OBSTACLES

The obstacles to any digital transformation are your dinosaurs, which we learned to identify in Chapter 2. The technological dinosaurs and operational dinosaurs are actually easy to deal with: They will be identified and dealt with as part of your well-thought-out and comprehensive plan.

But your human dinosaurs may take a bit more work.

Expect Resistance

Reluctance to move forward with change can come from any-where (and, most likely, everywhere) in your organization. The executive team will be resistant, looking at their existing invest-ment in infrastructure and fearful of trying something new and different. And the IT team—those dragon-keepers, so protec-tive of their systems as well as worried about their own jobs—may be reluctant to support any change at all.

That leaves the success of digital transformation largely in the hands of the new IT leader, which is why the IT leader's role *has to* drastically change, as we talked about in Chapter 5. Of course, the CDO still needs to understand technology! But the role of the IT leader is increasingly about coming up with a vision for a new and better future, and about communicating that vision.

The new CDO or IT leader must be able to build a shared vision with their peers in the leadership team and ensure buy-in across the organization. They need support at all levels, from the CEO out to the people working in the field.

They need to convince the CEO and the board that the digital transformation plan is vital for the company's survival, because the IT leader needs their support—both in principle and finan-cially. This may not be easy if there are some dinosaurs at the corporate or board level who are not capable of comprehending technological change, or who simply refuse to.

As the IT leader's role has changed, the roles of *all* C-suite
executives have begun to change as well. Part of every executive's job now is to be aware of and understand new and
emerging technologies. Of course, some executives and board
members may be uncomfortable operating outside their own
comfort zone, so the new IT leader must provide opportunities for them to learn and understand the changes. Additionally,
there may need to be difficult discussions and perhaps difficult
decisions regarding those who cannot or will not keep pace
with change.

At the same time, don't overlook the new opportunities opening up; there may be scope for the creation of new C-suite roles
as well. In addition to the chief digital officer or CDO, it might
serve some organizations to install new positions such as chief
automation officer or chief information security officer.

Employees at all levels need to be reassured that they won't
lose their jobs. Your plan needs to account for their continued
employment, be that through re-skilling and shuffling personnel
to different roles in-house, or by outsourcing them to your new
tech partners.

Many corporations resist strategic outsourcing due to the
significant organizational upheaval that it creates. However, in
today's world and with business operations needing to be cloud-
based, strategic outsourcing of legacy operations is an essential
step to any modernization plan. The new IT leader understands

this and is committed to finding the appropriate tech partner(s) or systems integrators to lead them on this journey.

THE VISIONARY LEADER

There are many points that I am trying to make in this book—all of which I deem important. But if I were forced to reduce my message to only a single point, it would be this: The role of the IT leader has changed. The IT leader is no longer the engineer down in the basement, keeping the lights on and setting policy on passwords. The new CDO is both a visionary and a communicator.

How to Be a Visionary

A visionary is someone with a strong vision of the future. Be they right, be they wrong—they have a path which guides them, and which helps them to lead with a specific end point in mind: their unique vision of what is to come.

"I very rarely get pulled into the today," Jeff Bezos, the founder of the world's second-largest company (by revenue), Amazon, told *Forbes* in 2018. "I get to work two or three years into the future, and most of my leadership team has the same setup."[3]

---

3   Randall Lane, "Bezos Unbound: Exclusive Interview with the Amazon Founder on What He Plans to Conquer Next," *Forbes*, September 4, 2018, https://www.forbes.com/sites/randalllane/2018/08/30/bezos-unbound-...

Who are the great visionaries of our time? We think of people like Bezos, or names like Bill Gates and Steve Jobs, or Richard Branson and Elon Musk. These leaders' success is evidenced by their status, not only as leaders of major global corporations, but some of the richest people who have ever lived on our planet.

But how did they become visionaries? Were they born with these supernatural powers? Or is it something they learned, or practiced? I believe that they are in fact just regular people who, like Bezos, have made it their job to work two or three years into the future.

I think of the path to becoming a world-class visionary as parallel to that of becoming a world-class tennis player. Yes, of course, some good genetics help the cause: a bit of extra brain-power or muscle, depending upon what your goal is. But good genetics does nothing if you don't put in the hard work.

Imagine an aspiring tennis player who has all the great genetics—they were born with the right fast-twitch muscles, and powerful legs, and the right neural connections for lightning-quick hand-eye coordination—but they never get out on the court to train. How far in the tennis world will their superlative genetics ever take them?

Just like that gifted athlete who still needs to put in the time on the court and train, you need to invest in yourself: looking

...exclusive-interview-with-the-amazon-founder-on-what-he-plans-to-conquer-next/#5e3c64c5647b.

ahead at what may come, understanding the issues and the variables, and thinking about how to deal with what you see. Or even better—how to take *advantage* of what you see.

That is what a visionary does: They look ahead and they anticipate. They predict, they forecast! Being a visionary means expecting and being *proactive*, rather than monitoring what is happening and being *reactive*.

Here is a list of a few of the fields that I keep an eye on at present, areas that are rapidly evolving and currently changing the world we live in. But be aware. If you are a *true* visionary (or on the road to becoming one), you know that change is constant, and therefore that this list will change and evolve too.

- Energy innovation: alternative sources of energy, as well as changes in how energy is distributed and stored.
- AI: Yes, we have already devoted two chapters to AI. But by the time these words are in print, the AI world will have changed. So, keep an eye on it!
- Robotics: Far beyond our old ideas of robots manufactured in human form, robots today range from submillimeter in size to huge machines. Currently mostly applied to automate industrial processes, robots of the future will rely increasingly on AI to sense their surroundings and make decisions, becoming ever more humanlike.

- Autonomous vehicles: Another natural combination of AI and robotics, driverless vehicles will be used for everything from your daily commute to work to airborne deliveries and more.
- Quantum computing: Based on atoms and subatomic particles rather than digital bits, quantum computing will result in exponentially faster computer solutions, which is especially relevant to power the complex needs of AI.
- Enhanced reality: Once mainly the realm of computer games, enhanced reality and its variations (augmented reality, virtual reality, and mixed reality versions) will have increased applicability to fields such as engineering, education, and entertainment.
- Transhumanism: Implantation of devices into humans will range from those that undertake discrete functions such as opening doors or making payments, to enabling new "senses" such as the ability to detect direction or augment color vision.
- A final factor—although not exactly a technology—that you need to pay attention to is geopolitical forces.

Some of these geopolitical forces are "soft," meaning they are powered by a gentle swing in consumer sentiment. An example would be the growing public acceptance of climate change and the science behind it. This may drive investment in technologies as diverse as lab-grown meat, alternative energy sources, and high-tech solutions for dealing with plastic waste. It may also drive investment away from some technologies. For example, increasing awareness of problems with plastic pollution in the oceans and microplastics has spurred trends such as returning to natural fibers such as wool for clothing.

Other geopolitical forces are "hard" and much more in-your-face, such as war. The 2022 invasion of Ukraine by Russia quickly changed the game in many industries. For example, Elon Musk's SpaceX satellite system suddenly had an important tactical role to play in Ukraine communications on and from the battlefields. Additionally, many European countries abruptly became pressured to end their reliance on Russian gas and accelerate development of more local energy sources or alternative energy technologies. The world continues to be a place of rapid change.

INSPIRING YOUR TEAM

No matter how prescient a business executive may be, they will get nowhere if they cannot bring their team on board. The first part of becoming a successful visionary is the ability to envision an improved or expanded place for your organization in the future. But the second critical part of being a visionary *leader* is the actual leading. They need to be able to communicate that vision and rally their team members to support it.

Steve Jobs was a master in this department. Yes, he was responsible for envisioning an amazing new future, populated with a range of devices that have changed not only how business is conducted, but how we go about our lives. But critically, he was able to rally and inspire his team to get behind him and support that vision—as well as inspire all of us to want his products.

That is what I endeavored to do with initiatives such as my "Jaguar E-type" metaphor at National Geographic. I was far from the first person to arrive at Nat Geo with a vision for change! Leaders before me had also had a vision of what was needed to bring that organization into the future. However, they were not able to rally their team to change direction: to stop what they were doing and to be convinced that this new way would bring them something better.

This is my message, and I'll say it again: The role of the IT leader indeed has changed. All the tech smarts in the world will not bring about change in your company. What is required is being a leader and finding a way in which you can lead and which your people want to follow.

CONTINUAL EVOLUTION

How will you know when your job is done?

I am sorry to break it to you. The job will never be done. Back to what I just said above: Embrace change. Because there will always be more meteors coming.

The CEOs and executive leaders of the future will be those who truly understand the power of technology to transform their organizations—but who also realize that digital transformation is about change (and constant change), not just technology.

You will succeed by framing transformation as a major business initiative that touches *all* parts of the organization. Digital transformation cannot succeed if it is viewed as a simple project, or as part of a "business as usual" technology rollout. In communicating with your team, emphasize the "transformation" rather than the "digital." That way, you acknowledge that the initiatives will make a measurable and positive impact on the organization: They will *transform* how you all do business.

Remember, you are changing culture and people, not just computer systems. Your digital transformation means a continual evolution in how interactions take place—yes, of course, in how your people interact with technology, but also in how your people interact with other people and how your technology interacts with technology. And how your people and your technology all interact with and respond to external factors: those new meteors that will only keep coming.

Acknowledgements

To my extraordinary wife, Alina—thank you for your unwavering patience and boundless support, and for always reminding me that even the most profound transformations begin with a single spark. Your belief in me made this possible. This book is for you—and for our amazing family, whose love fuels everything I do.

And to the visionary leaders who shaped my journey: Peter Breakell, Val Feerick, Will Grannis, Geoff Henderson, Max Mancini, John Tadman, Rachel Webber, and many others. Your insight, courage, and refusal to settle for the status quo profoundly influenced how I think about innovation and transformation. I am deeply grateful for your mentorship and the investment you made in my growth.

I wouldn't be the leader I am today without you.

About the Author

Marcus East is a globally recognized technology executive, investor, and advisor, known for driving large-scale digital innovation and significant business value. His unique perspective stems from senior leadership roles at both pioneering tech giants (Apple, Google, IBM) and iconic brands undergoing digital transformation (National Geographic, T-Mobile).

With deep expertise in digital strategy, customer-facing technology, and AI-driven innovation, Marcus consistently tackles complex challenges. He previously spearheaded T-Mobile's digital strategy and led National Geographic's successful evolution into a top social media brand. As part of Google's Office of the CTO, he advised executives on transformative AI solutions.

Beyond his executive career, Marcus is an active investor and board member for high-growth, AI-driven start-ups, particularly in health care and media. Marcus serves on the Board of

Directors at Seattle Children's Hospital—widely recognized as one of the nation's leading pediatric institutions—where he brings his deep commitment to innovation and the transformative power of technology. He is particularly focused on how emerging technologies can improve healthcare delivery, enhance patient outcomes, and help address systemic challenges in pediatric care.

A fellow of the British Computer Society and a member of the Chartered Management Institute, Marcus is passionate about leveraging technology for positive social change. He holds dual American and British citizenship and splits his time between London and New York.

www.ingramcontent.com/pod-product-compliance
Lightning Source LLC
Chambersburg PA
CBHW071208210326
41597CB00016B/1725